RISK-BASED WASTE CLASSIFICATION IN CALIFORNIA

COMMITTEE ON RISK-BASED CRITERIA FOR
NON-RCRA HAZARDOUS WASTE

BOARD ON ENVIRONMENTAL STUDIES AND TOXICOLOGY

COMMISSION ON LIFE SCIENCES

NATIONAL RESEARCH COUNCIL

NATIONAL ACADEMY PRESS
WASHINGTON, D.C.

NATIONAL ACADEMY PRESS   2101 Constitution Ave., N.W.   Washington, D.C. 20418

NOTICE: The project that is the subject of this report was approved by the Governing Board of the National Research Council, whose members are drawn from the councils of the National Academy of Sciences, the National Academy of Engineering, and the Institute of Medicine. The members of the committee responsible for the report were chosen for their special competences and with regard for appropriate balance.

The National Academy of Sciences is a private, nonprofit, self-perpetuating society of distinguished scholars engaged in scientific and engineering research, dedicated to the furtherance of science and technology and to their use for the general welfare. Upon the authority of the charter granted to it by the Congress in 1863, the Academy has a mandate that requires it to advise the federal government on scientific and technical matters. Dr. Bruce Alberts is president of the National Academy of Sciences.

The National Academy of Engineering was established in 1964, under the charter of the National Academy of Sciences, as a parallel organization of outstanding engineers. It is autonomous in its administration and in the selection of its members, sharing with the National Academy of Sciences the responsibility for advising the federal government. The National Academy of Engineering also sponsors engineering programs aimed at meeting national needs, encourages education and research, and recognizes the superior achievements of engineers. Dr. William A. Wulf is president of the National Academy of Engineering.

The Institute of Medicine was established in 1970 by the National Academy of Sciences to secure the services of eminent members of appropriate professions in the examination of policy matters pertaining to the health of the public. The Institute acts under the responsibility given to the National Academy of Sciences by its congressional charter to be an adviser to the federal government and, upon its own initiative, to identify issues of medical care, research, and education. Dr. Kenneth I. Shine is president of the Institute of Medicine.

The National Research Council was organized by the National Academy of Sciences in 1916 to associate the broad community of science and technology with the Academy's purposes of furthering knowledge and advising the federal government. Functioning in accordance with general policies determined by the Academy, the Council has become the principal operating agency of both the National Academy of Sciences and the National Academy of Engineering in providing services to the government, the public, and the scientific and engineering communities. The Council is administered jointly by both Academies and the Institute of Medicine. Dr. Bruce M. Alberts and Dr. William A. Wulf are chairman and vice chairman, respectively, of the National Research Council.

This project was supported by Research Agreement No. 98RA1539 between the National Academy of Sciences and the University of California. Any opinions, findings, conclusions, or recommendations expressed in this publication are those of the author(s) and do not necessarily reflect the view of the organizations or agencies that provided support for this project.

International Standard Book Number 0-309-06544-5

Additional copies of this report are available from:

National Academy Press
2101 Constitution Ave., NW
Box 285
Washington, DC 20055
800-624-6242
202-334-3313 (in the Washington metropolitan area)
http://www.nap.edu

Copyright 1999 by the National Academy of Sciences. All rights reserved.

COMMITTEE ON RISK-BASED CRITERIA FOR NON-RCRA HAZARDOUS WASTE

ROGENE HENDERSON *(Chair)*, Lovelace Respiratory Research Institute, Albuquerque, New Mexico
MARK W. BELL, Parsons Brinckerhoff Energy Services, Denver, Colorado
JOSEPH F. BORZELLECA, Virginia Commonwealth University, Richmond, Virginia
EDWIN H. CLARK, II, Clean Sites, Washington, DC
EDMUND A.C. CROUCH, Cambridge Environmental Inc., Cambridge, Massachusetts
JOHN P. GIESY, Michigan State University, East Lansing, Michigan
P. BARRY RYAN, Emory University, Atlanta, Georgia
JAMES N. SEIBER, U.S. Department of Agriculture, Albany, California
CURTIS C. TRAVIS, Project Performance Corporation, Knoxville, Tennessee

Staff

RAYMOND A. WASSEL, Program Director
ROBERTA WEDGE, Staff Officer
ROBERT J. CROSSGROVE, Editor
MIRSADA KARALIC-LONCAREVIC, Information Specialist
RUTH DANOFF, Senior Program Assistant
LUCY V. FUSCO, Project Assistant

Sponsor

Department of Toxic Substances Control,
California Environmental Protection Agency

BOARD ON ENVIRONMENTAL STUDIES AND TOXICOLOGY

GORDON ORIANS (*Chair*), University of Washington, Seattle, Washington
DONALD MATTISON (*Vice Chair*), March of Dimes, White Plains, New York
DAVID ALLEN, University of Texas, Austin, Texas
MAY R. BERENBAUM, University of Illinois, Urbana, Illinois
EULA BINGHAM, University of Cincinnati, Cincinnati, Ohio
PAUL BUSCH, Malcolm Pirnie, Inc., White Plains, New York
PETER L. DEFUR, Virginia Commonwealth University, Richmond, Virginia
DAVID L. EATON, University of Washington, Seattle, Washington
ROBERT A. FROSCH, Harvard University, Cambridge, Massachusetts
JOHN GERHART, University of California, Berkeley, California
MARK HARWELL, University of Miami, Miami, Florida
ROGENE HENDERSON, Lovelace Respiratory Research Institute, Albuquerque, New Mexico
CAROL HENRY, Chemical Manufacturers Association, Arlington, Virginia
BARBARA HULKA, University of North Carolina, Chapel Hill, North Carolina
DANIEL KREWSKI, Health Canada and University of Ottawa, Ottawa, Ontario
JAMES A. MACMAHON, Utah State University, Logan, Utah
MARIO J. MOLINA, Massachusetts Institute of Technology, Cambridge, Massachusetts
CHARLES O'MELIA, Johns Hopkins University, Baltimore, Maryland
KIRK SMITH, University of California, Berkeley, California
MARGARET STRAND, Oppenheimer Wolff Donnelly & Bayh, LLP, Washington, D.C.
TERRY F. YOSIE, Chemical Manufacturers Association, Arlington, Virginia

Senior Staff

JAMES J. REISA, Director
DAVID J. POLICANSKY, Associate Director and Senior Program Director for Applied Ecology
CAROL A. MACZKA, Senior Program Director for Toxicology and Risk Assessment
RAYMOND A. WASSEL, Senior Program Director for Environmental Sciences and Engineering
KULBIR BAKSHI, Program Director for the Committee on Toxicology
LEE R. PAULSON, Program Director for Resource Management

COMMISSION ON LIFE SCIENCES

MICHAEL T. CLEGG *(Chair)*, University of California, Riverside, California
PAUL BERG *(Vice Chair)*, Stanford University, Stanford, California
FREDERICK R. ANDERSON, Cadwalader, Wickersham & Taft, Washington, D.C.
JOHN C. BAILAR III, University of Chicago, Chicago, Illinois
JOANNA BURGER, Rutgers University, Piscataway, New Jersey
SHARON L. DUNWOODY, University of Wisconsin, Madison, Wisconsin
DAVID EISENBERG, University of California, Los Angeles, California
JOHN EMMERSON, Portland, Oregon
NEAL FIRST, University of Wisconsin, Madison, Wisconsin
DAVID J. GALAS, Keck Graduate Institute of Applied Science, Claremont, California
DAVID V. GOEDDEL, Tularik, Inc., South San Francisco, California
ARTURO GOMEZ-POMPA, University of California, Riverside, California
COREY S. GOODMAN, University of California, Berkeley, California
HENRY HEIKKINEN, University of Northern Colorado, Greeley, Colorado
BARBARA S. HULKA, University of North Carolina, Chapel Hill, North Carolina
HANS J. KENDE, Michigan State University, East Lansing, Michigan
CYNTHIA KENYON, University of California, San Francisco, California
MARGARET G. KIDWELL, University of Arizona, Tucson, Arizona
BRUCE R. LEVIN, Emory University, Atlanta, Georgia
OLGA F. LINARES, Smithsonian Tropical Research Institute, Miami, Florida
DAVID LIVINGSTON, Dana-Farber Cancer Institute, Boston, Massachusetts
DONALD R. MATTISON, March of Dimes, White Plains, New York
ELLIOT M. MEYEROWITZ, California Institute of Technology, Pasadena, California
ROBERT T. PAINE, University of Washington, Seattle, Washington
RONALD R. SEDEROFF, North Carolina State University, Raleigh, North Carolina
ROBERT R. SOKAL, State University of New York, Stony Brook, New York
CHARLES F. STEVENS, The Salk Institute, La Jolla, California
SHIRLEY M. TILGHMAN, Princeton University, Princeton, New Jersey
JOHN L. VANDEBERG, Southwest Foundation for Biomedical Research, San Antonio, Texas
RAYMOND L. WHITE, University of Utah, Salt Lake City, Utah

WARREN R. MUIR, Executive Director

OTHER REPORTS OF THE BOARD ON ENVIRONMENTAL STUDIES AND TOXICOLOGY

Research Priorities for Airborne Particulate Matter: I. Immediate Priorities and a Long-Range Research Portfolio (1998)
The National Research Council's Committee on Toxicology: The First 50 Years (1997)
Toxicologic Assessment of the Army's Zinc Cadmium Sulfide Dispersion Tests (1997)
Carcinogens and Anticarcinogens in the Human Diet: A Comparison of Naturally Occurring and Synthetic Substances (1996)
Upstream: Salmon and Society in the Pacific Northwest (1996)
Science and the Endangered Species Act (1995)
Wetlands: Characteristics and Boundaries (1995)
Biologic Markers [Urinary Toxicology (1995), Immunotoxicology (1992), Environmental Neurotoxicology (1992), Pulmonary Toxicology (1989), Reproductive Toxicology (1989)]
Review of EPA's Environmental Monitoring and Assessment Program (three reports, 1994-1995)
Science and Judgment in Risk Assessment (1994)
Ranking Hazardous Waste Sites for Remedial Action (1994)
Pesticides in the Diets of Infants and Children (1993)
Issues in Risk Assessment (1993)
Setting Priorities for Land Conservation (1993)
Protecting Visibility in National Parks and Wilderness Areas (1993)
Dolphins and the Tuna Industry (1992)
Hazardous Materials on the Public Lands (1992)
Science and the National Parks (1992)
Animals as Sentinels of Environmental Health Hazards (1991)
Assessment of the U.S. Outer Continental Shelf Environmental Studies Program, Volumes I-IV (1991-1993)
Human Exposure Assessment for Airborne Pollutants (1991)
Monitoring Human Tissues for Toxic Substances (1991)
Rethinking the Ozone Problem in Urban and Regional Air Pollution (1991)
Decline of the Sea Turtles (1990)
Tracking Toxic Substances at Industrial Facilities (1990)

Copies of these reports may be ordered from
the National Academy Press
(800) 624-6242 or (202) 334-3313
www.nap.edu

Preface

THE DEPARTMENT OF TOXIC SUBSTANCES CONTROL (DTSC) of the State of California Environmental Protection Agency is in the process of complying with the Regulatory Structure Update. The Regulatory Structure Update is a comprehensive review and refocusing of California's system for identifying and regulating management of hazardous wastes. As part of this effort, the DTSC proposes to change its current waste classification system that categorizes wastes as hazardous or nonhazardous based on their toxicity. Under the proposed system there would be two risk-based thresholds rather than the single toxicity threshold currently used to distinguish between the wastes. Wastes that contain specific chemicals at concentrations that exceed the upper threshold will be designated as hazardous; those below the lower threshold will be nonhazardous; and those with chemical concentrations between the two thresholds will be "special" wastes and subject to variances for management and disposal. The proposed DTSC system combines toxicity information with short or long-term exposure information to determine the risks associated with the chemicals.

Under section 57004 of the California Health and Safety Code, the scientific basis of the proposed waste classification system is subject to external scientific peer review by the National Academy of Sciences, the University of California, or other similar institution of higher learning or group of scientists. This report addresses that regulatory requirement.

This report is intended to assist the DTSC in determining whether the proposed waste classification method will be scientifically well grounded and protective of human health and the environment. It identifies areas where the proposed DTSC approach is appropriate, as well as areas where the DTSC documentation is not sufficient. The report indicates where and what type of improvements could be made to clarify both the presentation of the approach and the goals of the classification system.

To prepare the report, the committee reviewed the materials supplied to it by the DTSC, additional supporting materials received from DTSC and other individuals and organizations, and the information gathered at two public meetings held in Irvine, California on September 10, and November 20, 1998. The committee wishes to thank the following members of the California Environmental Protection Agency's Department of Toxic Substances Control: Jesse Huff, Director, Robert Borzelleri, Chief Deputy Director, Dr. Robert Stephens, Deputy Director, David Nunenkamp, Daniel Weingarten, and Drs. James Carlisle, Edward Butler, Bart Simmons, and John Christopher; and Dr. Lauren Zeise of the Office of Environmental Health Hazard Assessment for providing the committee with information on the models and framework for the risk methodology, for their presentations at the public meetings, and for their responses to written questions from committee members. We also gratefully acknowledge Barnes Johnson, U.S. Environmental Protection Agency, Office of Solid Waste and Emergency Response; Jane Williams, California Communities Against Toxics; Michael Lakin and Michael Easter, ICF Kaiser International (representing the Western States Petroleum Association); Linda M. Shandler and Brent C. Perry, County Sanitation Districts of Los Angeles County; Paul W. Abernathy, representing Mercury Technologies International/Advanced Environmental Recycling Corporation; Victor Hanna, City of Los Angeles, Bureau of Sanitation; Aspet Chater, Hugo Neu-Proler Company; Kirk T. Larson, representing San Diego Industrial Environmental Association; David Kay, Southern California Edison Company; and Charles A. White, Waste Management, Inc., for providing background information and for making presentations to the committee.

This report has been reviewed in draft form by individuals chosen for their diverse perspectives and technical expertise, in accordance with procedures for reviewing NRC reports approved by the NRC's Report Review Committee. The purpose of this independent review is to provide candid and critical comments that will assist the NRC in making the published report as sound as possible and to ensure that the report meets institutional standards for objectivity, evidence, and responsiveness to the study charge. The content of the final report is the responsibility of the NRC and the study committee, and not the responsibility of the reviewers. The review comments and draft manuscript remain confidential to protect the integrity of the deliberative process. We wish to thank the following individuals, who are neither officials nor employees of the NRC, for their participation in the review of this report: John C. Bailar, University of Chicago; Karen Florini, Environmental Defense Fund; Rolf Hartung, University of Michigan; Carol Henry, American Petroleum Institute; Donald M. Mackay, Trent University; Donald Mattison, March of Dimes Birth Defects Foundation; Glenn Paulson, Paulson and Cooper, Inc.; Emil Pfitzer, Ramsey, NJ; and Kenneth W. Sexton, University of Minnesota. These reviewers have provided many constructive comments and suggestions; it must be emphasized, however, that responsibility for the final content of this report rests entirely with the authoring committee and the NRC.

I am also grateful for the assistance of the NRC staff in the preparation of this report. In particular, the committee wishes to acknowledge Roberta Wedge, staff officer for the committee and Raymond A. Wassel, senior program director with the Board on Environmental Studies and Toxicology. Other staff members who contributed to this effort are Robert J. Crossgrove, editor; Ruth Danoff, senior program assistant; and Lucy V. Fusco, project assistant.

Finally, I would like to thank the members of the committee for their valuable expertise and dedicated efforts throughout the preparation of this report. Their efforts in preparing this report within a very short time frame are much appreciated.

> Rogene F. Henderson, Ph.D.
> *Chair*, Committee on Risk-Based
> Criteria for Non-RCRA Hazardous Waste

Contents

ABBREVIATIONS .... *xiii*

EXECUTIVE SUMMARY *1*

1 INTRODUCTION 14
  The Committee's Task and Aproach, *14*
  Report Organization, *16*
  Overview of California's Approach to Classifying Hazardous
      Waste, *17*

2 DTSC'S PROPOSED OVERALL APPROACH 36
  Statement of Goals, *38*
  Multimedia and Multipathway Risk Assessment, *41*
  Realistic Exposure Scenarios, *44*
  Valid Science, *45*
  Transparency, *48*
  Flexibility, *50*
  Implementation Practicality and Evaluation, *52*

3 SCENARIO SELECTION AND MODELING 55
  Exposure Scenarios: Purpose, *55*
  Modeling: Purpose, *56*
  Summary of DTSC Exposure Scenarios, *56*
  Modeling Used in the Scenarios, *58*

Analysis of Scenarios and Modeling, *60*
Ecological Scenario, *74*

4   ISSUES OF MODEL APPLICATION *77*
Model Parameters, *77*
Parameter Selection Within Specific Models, *83*
Analytical Methods, *102*
Toxicity Tests, *106*

5   MEETING PROGRAM GOALS *113*
RSU Guiding Principles, *114*
DTSC Program Goals, *116*
Other Considerations for DTSC's Approach, *119*
Program Evaluation, *121*

REFERENCES *123*

APPENDIX A: Biographical Information on the Committee on Risk-Based Criteria for Non-RCRA Hazardous Waste *126*

APPENDIX B: DTSC Issues *130*

APPENDIX C: List of Public Access Materials Received by the NRC Committee on Risk-Based Criteria for Non-RCRA Hazardous Waste *136*

APPENDIX D: Letter of Introduction, Overview, Concept Paper, and Appendices 1–4 from DTSC Report *145*

Abbreviations

ADOM	acid deposition and oxidant model
AWQC	ambient water quality criteria
Cal/EPA	California Environmental Protection Agency
CAM	California Assessment Manual
CCR	California Code of Regulations
CFR	Code of Federal Regulations
CUPA	Certified Unified Program Agencies
CV	coefficient of variation
DAF	dilution attenuation factor
DDT	dichloro diphenyl trichloroethane
DTSC	Department of Toxic Substances Control
EFH	Exposure Factors Handbook
EP	extraction procedure
EPA	U.S. Environmental Protection Agency
EQL	estimated quantitation level
FDM	fugitive dust model
HWIR	Hazardous Waste Identification Rule
ISC	industrial source complex dispersion model
IWMB	Integrated Waste Management Board
LC	lethal concentration
LD	lethal dose
LOD	limit of detection
LOQ	limit of quantitation

MCL	maximum contaminant level
MSWL	municipal solid waste leachate
NAAQS	National Ambient Air Quality Standards
NIOSH	National Institute for Occupational Safety and Health
NRC	National Research Council
OSHA	Occupational Safety and Health Administration
PEA	preliminary endangerment assessment (model)
RAGS	Risk Assessment Guidance for Superfund
RCRA	Resource Conservation and Recovery Act
RSU	regulatory structure update
SD	standard deviation
SERT	soluble or extractable regulatory threshold
STLC	soluble threshold limit concentration
SWRCB	State Water Resources Control Board
TCDD	2,3,7,8-tetrachloro-dibenzo-p-dioxin
TCE	trichloroethylene
TCLP	toxicity characteristic leaching procedure
TTLC	total threshold limit concentration
WET	waste extraction test

Risk-Based Waste Classification In California

Executive Summary

THE COMMITTEE'S TASK

IN THE STATE of California, the management of wastes considered potentially hazardous is first regulated by the state in compliance with the federal Resource Conservation and Recovery Act (RCRA). For wastes that are not regulated under RCRA as hazardous, the state uses an additional classification system to determine whether such wastes pose a threat to human health and the environment. This report focuses on a new approach proposed for use in California to classify wastes that are not considered to be hazardous wastes under RCRA.

As with RCRA, California classifies waste as hazardous or nonhazardous based on four characteristics: reactivity, ignitability, corrosivity, or toxicity. As part of its regulatory structure update process, the California Environmental Protection Agency's Department of Toxic Substances Control (DTSC) has reviewed those characteristics and has undertaken to propose a new waste-classification system for hazardous wastes.[1] The classification of a waste determines how the waste will be managed (e.g., storage, transport, disposal, reporting requirements).

DTSC has spent more than 2 years developing a new approach for the toxicity characteristic. The proposed system has two basic goals: (1)

---

[1] In this report, unless noted otherwise, the term "hazardous waste" refers to waste that would be classified as hazardous by the State of California's proposed revisions of its toxicity criteria.

to move from a classification system that is toxicity-based to a risk-based system that considers both toxicity and potential exposure to the waste[2] and (2) to replace the one-threshold system that classifies waste as either hazardous or nonhazardous with a two-threshold system that classifies waste incrementally as "hazardous" (high toxicity), "special" (moderate toxicity), or "nonhazardous." The committee notes that the terms hazardous waste and nonhazardous waste, while appropriate for a classification system based on toxicity, might not be technically appropriate for one based on risk, when the definition of the term "hazardous" refers to the toxicity of an agent and is independent of exposure.

Before the proposed classification system and subsequent regulations can go into effect, DTSC is required by California statute to "conduct an external scientific peer review of the scientific basis of any new rule" (California Health and Safety Code § 57004). Consequently, DTSC requested that the National Research Council (NRC) of the National Academies of Sciences and Engineering conduct such a review of its proposed risk-based approach for the toxicity characteristic used to classify waste and suggest areas where the approach might be improved. This report is the independent review of DTSC's proposed waste-classification system by the National Research Council's Committee on Risk-based Criteria for Non-RCRA Hazardous Waste.

WHAT IS BEING REVIEWED?

The technical basis for the proposed waste-classification system is documented in DTSC's "Risk-Based Criteria for Non-RCRA Hazardous Waste: A Report to the National Research Council Introducing Proposed Changes to the Definition of Hazardous Waste in the California Code of Regulations." An overview of this report is presented in Appendix D.

---

[2]In this report, hazard (toxicity) is defined as the determination of whether a waste or waste component is or is not causally linked to a particular health effect, and of the nature and strength of the evidence for such causation; exposure is a determination of the intensity, frequency, and duration of contact with a waste or its components by an individual or population; and risk combines the hazard and exposure assessments to estimate the probability of specific harm to an exposed individual or population.

This approximately 1,500-page document, which was submitted to the NRC before the first committee meeting, includes a synopsis of the proposed classification system and many of the background materials used to develop the classification criteria.

The committee also examined supplementary DTSC documents, including those found on DTSC's home page on the Internet. The committee requested clarification of numerous aspects of the DTSC documentation both through written responses from DTSC and through verbal responses to questions posed by the NRC committee members at two public meetings. At the public meetings, the committee also gathered information from waste managers, waste generators, environmental groups, and other interested individuals and organizations. Comments on DTSC's approach were given in oral presentations as well as in written materials. Thus, the scientific and technical bases of the proposed classification system that were reviewed by the committee included information received before and during the committee's deliberations. The proposed classification system reviewed in this report is unchanged from the descriptions in the original documentation received by the NRC, albeit with some clarification on specific issues from DTSC and other individuals and organizations.

DTSC'S CURRENT CLASSIFICATION SYSTEM

As presented in its report, DTSC currently uses a one-threshold system to classify wastes either as hazardous or nonhazardous based on the toxicity of the whole waste or the waste components.[3] Under the current system, DTSC has established eight toxicity criteria by which a waste might be considered hazardous:

- Federal definition of a hazardous waste according to its toxicity characteristic.
- Exceedence of total threshold limit concentrations or soluble threshold limit concentrations for specific chemicals.
- Exceedence of acute oral toxicity threshold.

---

[3]The current system also includes certain specified types of "special wastes" that might have hazardous characteristics, but are not necessarily required to be disposed to hazardous waste landfills.

- Exceedence of acute dermal toxicity threshold.
- Exceedence of acute inhalation toxicity threshold.
- Exceedence of acute aquatic toxicity threshold (fish).
- The presence of one or more of 16 named carcinogens above a threshold.
- Otherwise shown by experience or testing to be hazardous to public health or the environment (new threats).

For wastes that are not considered hazardous under RCRA, DTSC currently determines whether the waste contains any specific chemicals that exceed thresholds established for soluble threshold limit concentrations, or whether the total amounts of any specific chemicals exceed established total threshold limit concentrations. If the concentration of any of those chemicals exceeds the relevant threshold, the waste is classified as hazardous; otherwise, the waste is classified as nonhazardous. Soluble and total threshold limit concentrations are established to protect humans and other organisms from adverse effects following exposure to those chemicals through contact with groundwater or by any other route (e.g., soil, air, surface water, and food).

If a waste does not contain chemicals that exceed any soluble threshold limit concentrations or total threshold limit concentrations, it can still be classified as hazardous based on its acute toxicity (oral, dermal, inhalation, or aquatic) or the presence of one of 16 specific carcinogens. There are thresholds for both acute toxicity and the 16 specific carcinogens. The acute toxicity of a waste is determined either by testing the whole waste or by summing the acute toxicity values of the individual waste components. A waste also might be considered hazardous if data are available to suggest that a new chemical or route of exposure might pose a threat to humans or other organisms.

DTSC'S PROPOSED CLASSIFICATION SYSTEM

The proposed system will continue to classify wastes as hazardous or nonhazardous; however, it will also include a category of special wastes that are considered to be less than hazardous but still might cause adverse effects if disposed of as nonhazardous waste. Special wastes might be subject to less restrictive reporting and disposal requirements than hazardous wastes.

In the proposed system, DTSC will modify five of the eight toxicity criteria listed in this summary by establishing two thresholds for each of the following: total threshold limit concentrations or soluble threshold limit concentrations; acute oral toxicity; acute dermal toxicity; acute inhalation toxicity; and acute aquatic toxicity. One threshold will distinguish hazardous wastes from special wastes (upper threshold). The other will distinguish special wastes from nonhazardous wastes (lower threshold), as shown in Figure ES-1.

DTSC also proposes to replace the soluble threshold limit concentrations with a new category referred to as soluble or extractable regulatory thresholds, and to develop upper and lower thresholds for that category as well. The classification criteria for wastes containing carcinogens and for wastes posing new threats would remain the same as in the current system.

Unlike the present system, which classifies wastes based on their toxicity, the proposed system classifies wastes based on their potential risks to the health of humans and other organisms. The risk posed by the waste is estimated from (1) its long-term risk (chronic toxicity and long-term exposure potential), (2) its short-term risk (acute toxicity and short-term exposure potential), or (3) both its long-term and short-term risks. A variety of models are used to determine the exposure potentials for the chemicals in the waste.

To calculate the two new total threshold limit concentrations, DTSC used four scenarios by which humans or other organisms might be exposed to chemicals in the wastes: (1) residents living near a hazardous waste landfill; (2) waste workers at a landfill; (3) residents living on land to which waste had been previously applied; and (4) ecological (nonhuman) receptors. Those scenarios were designed to protect workers, residents, and ecological receptors from chronic adverse effects resulting from long-term exposures. Multipathway exposure models were used to estimate the risks associated with each of the human health scenarios. Risks to ecological receptors were based primarily on risk assessments developed in the technical support document for the U.S. Environmental Protection Agency's proposed Hazardous Waste Identification Rule.

The proposed upper and lower soluble or extractable regulatory thresholds were derived by a different method from that used to calculate the total threshold limit concentrations. For soluble or extractable regulatory thresholds, the only exposure scenario for humans was as-

Current System	Proposed System
Hazardous Waste*	Hazardous Waste
Toxicity-based Threshold	Risk-based Threshold (Upper)
Nonhazardous Waste	Special Waste
	Risk-based Threshold (Lower)
	Nonhazardous Waste

*Under the current system, a very small number of hazardous wastes receive variances for disposal as special wastes.

FIGURE ES-1 DTSC's waste-classification systems.

sumed to be ingestion of contaminated drinking water. Ecological receptors were assumed to be in direct contact with the groundwater.

For acute toxicity thresholds, acute exposure scenarios were used. The upper acute oral and dermal toxicity thresholds were based on exposure of adults, and the lower toxicity thresholds on exposure of children. For acute inhalation toxicity, upper and lower thresholds were developed separately for volatile chemicals and for particles. Acute toxicity values are applicable to whole wastes, either by direct testing of the whole waste or by appropriate summation of acute toxicity values for all the waste constituents.

In its report, DTSC also proposes to use the federal RCRA extraction test, the toxicity characteristic leaching procedure, rather than the waste extraction test, which it currently uses for extracting wastes to determine if the wastes have the potential for contaminating groundwater.

THE NRC COMMITTEE'S FINDINGS AND RECOMMENDATIONS

Improved flexibility in the management and disposal of wastes while protecting human health and the environment is a primary goal of the regulatory structure update process under which DTSC's proposed approach was developed. Flexibility is to be improved through the use of a two-threshold rather than a one-threshold classification system, a new risk-based procedure for developing soluble or extractable regulatory thresholds and total threshold limit concentrations for chemicals (those currently regulated as well as additional ones), the ability to incorporate new toxicity and exposure data for chemicals, and the potential to review the risks posed by entire wastes. The lower the risk posed by a waste, the greater the number of disposal options.

In the following sections, the committee discusses areas where DTSC's proposed approach is successful in meeting these goals, as well as areas where there are significant failings in either the scientific and technical aspects of the system or in the documentation of DTSC's approach. The committee makes recommendations for improving both the scientific underpinnings of the approach and its presentation.

Risk-Based Waste Classification

The committee concurs with DTSC's decision to develop a risk-based, multimedia, multipathway approach to classifying wastes. Such an approach considers not only the toxicity of the waste, but also the potential for people or other organisms to be exposed by multiple pathways in a variety of settings. In addition, the approach can be applied to protect humans and other organisms at levels of protection required by California policy. In order to illustrate the differences between the current and the proposed systems, the committee recommends that DTSC select several known wastes and show how they would be classified under each system.

Exposure Pathway Integration

The committee recommends that DTSC adopt a comprehensive approach to its proposed multimedia implementation by integrating risks from all exposure pathways into a single risk characterization. The exposure models used by DTSC should integrate all pathways because humans and other organisms might be exposed simultaneously through numerous pathways. This might require a slightly different approach by DTSC that would combine the results of the extraction tests (for groundwater protection) and the measurement of total concentrations in the waste.

Protection Goals and Exposure Scenarios

DTSC's proposed approach lacks clear, explicit definitions of the human and ecological populations or groups to be protected and the desired levels of protection for these populations, including sensitive subpopulations such as children. By linking human and ecological protection goals to specific exposure scenarios, the selection of appropriate parameters for the multimedia transport and exposure models would be improved.

The DTSC documentation should describe how all exposure scenarios were developed for residents, workers, and nonhuman organisms, and screened for acceptability and completeness. The rationale for choosing the long-term exposure scenarios for the total threshold limit concentrations and for the soluble or extractable regulatory thresholds should be clearly explained, as should the short-term exposure scenarios used to develop the acute toxicity thresholds.

DTSC should explain in its documentation how the model parameters were selected to ensure consistency with the appropriate exposure scenarios. A systematic approach is required to identify the appropriate parameters. Furthermore, DTSC should indicate how population or environmental shifts might be incorporated into future exposure scenarios. For example, how will changes in subpopulations, such as children, be factored into the scenarios?

Model Components and Parameters

The committee recommends that DTSC's proposed approach explicitly describe the physical processes presumed to occur in each environmental pathway of each exposure scenario, to allow determination of whether the assumptions are realistic.

In its review of the component models used in the exposure models,

EXECUTIVE SUMMARY

the committee found that specific models were not well defined, and that it was difficult or impossible to determine the original sources of many of the parameters used. The DTSC documentation should indicate how the specific models, with their chosen parameter values, match the physical processes occurring in each particular exposure scenario and what the original sources for all the parameter values were.

The committee found many errors and inconsistencies in: (1) the selection of models for the scenarios (application); (2) the documentation of the scenarios, models, and parameters; (3) the implementation of the models (i.e., what was actually done was not what DTSC says it did); and (4) the parameter values, which were often incorrect or mismatched to the scenarios. The committee recommends that a more thorough quality-control review and validation be conducted as these are critical for the development of a scientifically defensible classification system.

Model Performance

The DTSC proposal does not adequately distinguish between variability and uncertainty in its modeling efforts. With a more complete statement of policy goals, including the desired levels of protection and the populations to be protected, the correct treatment of uncertainty and variability should become apparent and should be explicitly discussed in the context of the policy goals.

The committee recommends that DTSC conduct a thorough sensitivity analysis to identify critical parameters and dominant pathways in the exposure models. Both the method for and results of the analysis should be clearly communicated, particularly given DTSC's statement that one exposure pathway dominated each scenario. The committee recommends that DTSC also reexamine the statistical distributions and determine what effect assumptions of lognormality have on the modeling results.

Analytical Issues

DTSC's proposal to use twice the estimated quantitation level, an analytical end point, when the calculated concentration of the risk-based soluble or extractable regulatory threshold is less than the estimated quantitation level, is not a risk-based approach. There is no connection between the sensitivity of an analytical method and the sensitivity of exposed organisms. The committee recommends that DTSC establish the protectiveness of this default value.

DTSC proposes to use the federal RCRA toxicity characteristic leaching procedure to extract wastes rather than the current California waste extraction test to detect soluble chemicals in landfill leachate. The committee finds that DTSC is generally on the right path, but recommends that DTSC work more closely with stakeholders and the U.S. Environmental Protection Agency to address the shortcomings of the toxicity characteristic leaching procedure and the waste extraction test, and to explore development of a new extraction test that does not have the shortcomings of these tests, before final action is taken.

Risk for Entire Waste

In DTSC's proposed approach, the chronic-risk estimate for a waste is based only on those chemicals for which total threshold limit concentrations or soluble or extractable regulatory thresholds have been developed. Chronic risks posed by other components of the waste or combinations of components are not considered, even though chronic risks from a combination of chemicals are likely to be greater than from any individual chemical. The committee, therefore, recommends that DTSC's proposed approach consider the chronic risk of all waste components, including those chemicals with total threshold limit concentrations or soluble or extractable regulatory thresholds, when assessing the total risk posed by the waste.

The committee also recommends that the DTSC approach take into account metal speciation, as the toxicity of a metal might depend greatly on the metal species present in the waste.

Chronic Toxicity Data and Irritation Testing

For wastes that do not contain chemicals for which a total threshold limit concentration or soluble or extractable regulatory threshold has been established, DTSC proposes that the toxicity characteristic for the waste

EXECUTIVE SUMMARY

would be based on risks from its acute toxicity. The committee recommends that chronic toxicity data, when available, be considered by DTSC when assessing the risks posed by a waste. If only acute toxicity data are available for the risk assessment, the committee recommends that DTSC follow standard practice and use an uncertainty factor to account for the lack of chronic toxicity data. Failure to consider chronic effects could seriously underestimate the risks posed to human health and the environment from long-term exposure to wastes.

The committee further recommends that DTSC consider including acute irritation and sensitization (allergenicity) testing in its proposed approach. Respiratory irritation can exacerbate existing health conditions, such as asthma, and repeated exposures to some compounds can lead to sensitization of some people.

Transparency of the Proposed Approach

Clear documentation of the proposed DTSC approach is vital. The report reviewed by the committee lacks a roadmap to the documentation (almost 1,500 pages) and a synthesis into a manageable form. The documentation should parallel the process used by DTSC in developing its risk-based approach, beginning with a comprehensive discussion of the California human and ecological populations to be protected and at what levels, and followed by a description of all exposure scenarios and an explanation of the rationale for selecting representative scenarios, model selection and modifications, and parameter values. DTSC should include in its documentation the following: (1) background materials, including unpublished materials; (2) all implementations (in this case, spreadsheets); and (3) adequate references for all data sources. The committee believes that the lack of clarity in DTSC's documentation, including a clear description of the actual process by which a waste is classified, is a serious failing.

Flexibility Goals

The committee recommends that DTSC reexamine its claim of improved flexibility of the proposed approach. DTSC does not specify in its documentation how reclassifying some wastes into new categories will reduce the regulatory or economic burdens or increase flexibility with regard to waste disposal while still being protective of humans and other organisms. It behooves DTSC to propose a feedback or adaptive management mechanism to ensure that human health and the environment are protected at the desired levels.

The committee further recommends that DTSC's proposed approach include a clear process by which (1) additional chemicals can be evaluated for inclusion on the list of regulated agents, and (2) regulated chemicals can be reevaluated for possible reassignment of thresholds or even deletion from the state's list. Such a process must include a mechanism for response to new information as it becomes available. The inclusion of several examples using chemicals with widely varying characteristics (e.g., organic or inorganic, hydrophobic or hydrophilic) or new data (e.g., new toxicity studies, information on persistence in the environment) would considerably increase public confidence in the utility of the new waste-classification system.

CONCLUSIONS

In principle, the committee agrees that the use of a risk-based multimedia approach for classifying wastes is appropriate. In its assessment of the proposed DTSC waste-classification system, the committee concluded that, at the most fundamental level, DTSC can improve its approach by (1) providing a clear statement of its protection goals and their regulatory context; (2) explicitly defining and justifying the exposure scenarios; and (3) applying these scenarios consistently throughout the subsequent modeling. Furthermore, DTSC should conduct a much more extensive and detailed internal review of its proposed classification process to ensure scientific and technical accuracy and clarity.

Although the committee has identified a number of specific technical concerns with regard to DTSC's classification proposal, these concerns should not be interpreted to suggest that the approach has insurmountable flaws. The committee believes that DTSC's basic approach of considering the likely risks associated with waste disposal rather than just the toxicity of the waste, and using multimedia, multipathway exposure models for assessing these risks, represents a significant improvement in developing waste management regulations. The committee appreciates the difficulty faced by DTSC, or any organization, when

EXECUTIVE SUMMARY 13

developing a risk-based approach to waste classification that has the flexibility to incorporate new scientific knowledge. The committee's review suggests that the proposed approach would benefit from additional development and thorough documentation. When fully developed and documented, DTSC's approach could be a useful tool for other states and the U.S. Environmental Protection Agency in addressing similar waste classification issues. The committee, therefore, urges DTSC to address the concerns presented in this report, and submit the revised report to a detailed and comprehensive scientific (and editorial) evaluation by California Environmental Protection Agency staff and other experts before any further external review.

1

Introduction

IN THIS CHAPTER, the NRC committee identifies the task that was given to the committee for the review of the California Environmental Protection Agency (Cal/EPA) Department of Toxic Substances Control's (DTSC's) proposed risk-based classification system for wastes that are not subject to the federal Resource Conservation and Recovery Act (RCRA). The committee's approach to its task and the materials that were reviewed are discussed. The chapter then indicates those areas of waste management that the committee considered to be beyond the scope of its task. A description of the organization of the body of the report and the topics that are covered in the remaining chapters are then presented. The remainder of the chapter is an overview of DTSC's current and proposed waste classification systems.

THE COMMITTEE'S TASK AND APPROACH

The NRC Committee on Risk-based Criteria for Non-RCRA Hazardous Waste was given the following task:

> The committee will review the scientific and technical aspects of the California Environmental Protection Agency's (Cal/EPA's) proposed approach for the classification of hazardous wastes that are not subject to the Federal Resource Conservation and Recovery Act (RCRA) program. The committee will consider the analytical and

INTRODUCTION

bioassay procedures developed for the proposed approach. In addition, it will consider the toxicological risk implications of the criteria proposed by Cal/EPA for classifying waste into three categories. The committee will assess the scientific framework for the implementation of the proposed regulatory approach and it will identify any potential improvements that can be made to the scientific techniques used in the proposed system.

DTSC provided the committee with a document entitled "Risk Based Criteria for Non-RCRA Hazardous Waste: A Report to the National Research Council Introducing Proposed Changes to the Definition of Hazardous Waste in the California Code of Regulations" (DTSC 1998a). That document contains the specifics of the DTSC approach for the development of new risk-based values for toxicity threshold limit concentrations (TTLCs) and soluble or extractable regulatory thresholds (SERTs), including the documentation for the CalTOX model and its modifications, the lead risk-assessment spreadsheet, the modified preliminary endangerment assessment model, the ecological risk assessments, and several sections on the use of the federal toxicity characteristic leaching procedure in place of the California waste extraction test. This 1,494-page document and the spreadsheets implementing the models were the bases for the committee's review of DTSC's proposed methodology. In addition, the committee made use of several publicly available documents on the DTSC's home page on the Internet (http://www.dtcs.ca.gov) that described the regulatory context in which the proposed waste classification was developed (DTSC 1998b). After a preliminary review of the DTSC report, the committee determined that additional clarification was required on specific issues, such as the scenarios that DTSC used to develop their upper and lower toxicity criteria. The committee prepared and submitted 86 questions to DTSC to clarify these issues. DTSC provided answers to each of the questions (DTSC, personal commun., October 9, 1998, see Appendix C, No. 28).

In addition to the DTSC report and ancillary materials, the committee held two public meetings (on September 10, 1998 and November 20, 1998) to gather information from DTSC and other interested parties. Before, during, and immediately after the public meetings, numerous comments were received from various industry representatives and waste generators, environmental and public interest groups, local waste handlers (e.g., sanitation facilities), and affected trade associations (e.g., scrap-recycling industry). DTSC also provided some supplemental materials to the committee (DTSC, personal commun., October 9, 1998, see Appendix C). A list of all materials reviewed by the committee is provided in Appendix C.

The regulatory, legal, and policy context within which DTSC is proposing to change its waste classification scheme is not considered to be within the scope of this report. Other issues not considered in this report include DTSC's authority under various California statutes; the list of chemicals for which DTSC has proposed TTLCs or SERTs; the economics of compliance with this classification method; ramifications for other California regulatory agencies; and the appropriateness of disposal requirements, predisposal management requirements, or regulation of discharges to sanitary sewers.

Instead, the focus of this report is on the scientific validity of DTSC's waste classification methodology. The management of a waste after it is classified as hazardous, nonhazardous, or special is outside the committee's task and was not included in the methodology provided by DTSC. When appropriate, however, the committee indicates where implementing the DTSC proposal might be difficult in terms of management of the various waste streams or where regulatory or management decisions that affect risk estimates might require comment. The committee notes that the terms hazardous waste and nonhazardous waste, while appropriate for a classification system based on toxicity, may not be technically appropriate for one based on risk, when the definition of the term "hazardous" refers to the toxicity of an agent and is independent of exposure.

REPORT ORGANIZATION

This report is arranged to reflect the committee's analysis of the proposed DTSC risk-based method for classifying hazardous waste. In Chapter 2, the overall approach of the DTSC to identification of hazardous waste using a risk-based rather than a hazard-based process is discussed. Chapter 2 examines DTSC's proposed approach for protecting the California population and environment from risks due to exposure to hazardous substances, the use of technical data as support for policy decisions, and the use of a chemical-by-chemical approach for classifying wastes, with the uncertainties surrounding such an approach. In Chapter 3, the committee examines DTSC's method of selecting exposure scenarios and the application of the models to the selected scenarios. The

INTRODUCTION

validity of the scenarios and models is assessed in terms of their protectiveness of the health of the California population and environment. The details of the component-model input parameters, assumptions made in running the exposure models, who the models are intended to protect, and other health-related values are explored in Chapter 4. Chapter 4 also compares the use of the two analytical extraction methods for leachates from solid waste (toxicity characteristic leaching procedure (TCLP) and waste extraction test (WET)) and DTSC's use of acute toxicity data for classifying wastes. Finally, in Chapter 5, the committee identifies some programmatic concerns with the proposed DTSC method and suggests modifications of the DTSC approach to enhance clarity and effectiveness.

OVERVIEW OF CALIFORNIA'S APPROACH TO CLASSIFYING HAZARDOUS WASTE

In this section, the committee presents a synopsis of the DTSC's hazardous waste-classification system as presented to the committee (DTSC 1998a) and from other sources, such as "A Risk-Based Regulatory System: Revision of California Non-Resource Conservation and Recovery Act Waste Classification Regulation" (DTSC 1998b). This synopsis is intended to assist the reader in understanding the regulatory context in which DTSC has proposed changes to the hazardous waste classification system, and what those changes are. The committee has attempted to provide an accurate portrayal of DTSC's program based on the documentation provided by DTSC (DTSC 1998a), DTSC's written responses to questions from the committee (DTSC, personal commun., October 9, 1998, see Appendix C, No. 28), information gathered by the committee at two public meetings, and additional written materials from DTSC and the public; however, this synopsis is not intended to be a comprehensive or legal description of the current or proposed hazardous waste classification systems. For a further description of DTSC's proposed system, refer to the DTSC report.

Statutory Framework

In 1976, the U.S. Congress passed the Resource Conservation and Recovery Act (RCRA) (Public Law 94-580). RCRA gave the U.S. Environmental Protection Agency (EPA) the authority to control hazardous waste from generation to disposal. Under Subtitle C of RCRA, EPA regulates the management, storage, treatment, and disposal of certain hazardous wastes including disposal in subtitle C landfills. In subtitle D, RCRA also set forth a framework for the management of nonhazardous wastes, including disposal in subtitle D landfills.

Under RCRA, states have primary responsibility for the implementation of hazardous-waste regulations. State programs, authorized and approved by the EPA, must be at least equivalent to the federal regulations; however, such state programs may be more stringent or broader in scope than the federal program. Under the Hazardous Waste Control Act of 1972 (California Health and Safety Code, §25100 et seq.), which precedes the federal RCRA statute, the State of California mandated that DTSC develop regulations for the identification, management, and disposal of hazardous wastes. DTSC, after complying with RCRA requirements, uses an additional regulatory process to supplement the federal statute. Under California's regulations, certain wastes not classified as hazardous under RCRA may be classified and managed as hazardous waste if DTSC determines that they might adversely affect human health or the environment when disposed of as unregulated wastes.

Current Waste-Classification System

Like the federal RCRA, the DTSC waste-classification system identifies wastes as hazardous on the basis of their ignitability, corrosivity, reactivity, or toxicity. California uses the federal RCRA criteria for ignitability and reactivity. For corrosivity, it broadens the definition to include corrosive solids in addition to liquids. The California waste-classification system differs from the federal RCRA system in two primary ways. First, California has developed a list of 800 chemicals whose presence in a waste introduces a rebuttable presumption of hazard due to a potential characteristic of concern (corrosivity, reactivity, ignitability, and/or toxicity) annotated for each such chemical. The presumption can be rebutted by the testing the waste for the characteristics listed for the chemicals present in the waste. For toxicity, the relevant tests are acute oral, dermal, and inhalation toxicity, and the fish bioassay. These 800 chemicals are not discussed in the DTSC report (DTSC 1998a) and their designation and use were not evaluated by the committee.

Second, RCRA bases the toxicity of a waste on the results of a toxicity characteristic leaching procedure (TCLP) test, whereas California

INTRODUCTION

uses eight criteria to determine the toxicity of a waste, only the first of which is equivalent to the federal statute. The California criteria are the following:

- Federal definition as a hazardous waste.
- Exceedence of TTLCs or soluble threshold limit concentrations (STLCs) for specific chemicals.
- An acute oral toxicity less than 2,500 milligrams (mg) of waste per kilogram (kg) of body weight.
- An acute dermal toxicity less than 4,300 mg/kg of body weight.
- An acute inhalation toxicity less than 10,000 parts per million (ppm) in air.
- An acute aquatic toxicity, based on a fish bioassay, less than 500 mg/liter (L) of water.
- The presence of any one or more of 16 carcinogens in waste at a concentration exceeding 10 mg/kg.
- Shown by experience or testing to be hazardous to public health or the environment (new threats).

Under the current one-threshold system, DTSC classifies wastes as either hazardous or nonhazardous based on whether the toxicity of the waste or its constituents exceed one or more of the toxicity criteria; there is also a small category of hazardous wastes called "special" wastes for which DTSC has granted variances for disposal to specific nonhazardous landfills. Hazardous wastes must be disposed to landfills that have double composite liners and leachate collection systems (class I).[1] Nonhazardous wastes may be disposed to municipal solid-waste (class III) landfills and are not further regulated by DTSC. Special wastes may

---

[1] California has three classes of landfills (DTSC 1998b; DTSC personal commun., January 26, 1999; see Appendix C, No. 42):

Class I — DTSC landfills with the strictest controls available for hazardous waste, having a double composite liner and leachate collection system; similar to RCRA subtitle C landfill requirements.

Class II — Regional Waster Quality Control Boards landfills designed to protect waters of California by having a composite liner and leachate collection system; similar to RCRA subtitle D landfills with new liner requirements.

Class III — Integrated Waste Management Board landfills are the standard municipal-type of landfill; similar to RCRA subtitle D landfills without new liner requirements.

go to class I or non-class I landfills, such as class II, but must meet all other hazardous waste regulations before disposal.

DTSC distinguishes between wastes that are likely to leach into groundwater and those wastes for which exposure is from pathways other than groundwater. To determine whether a waste may be classified as hazardous due to the groundwater pathway, the waste is extracted using a California-specific test called the Waste Extraction Test (WET). The concentrations of specific chemicals in the extract are then compared with the STLCs to determine whether the waste is hazardous. STLCs are derived from the California drinking-water maximum contaminant levels (MCLs) or from fish toxicity values as designated in the EPA's ambient water quality criteria. STLCs have been established for 36 chemicals: the values of 13 are identical to the RCRA toxicity characteristic values, 4 are higher than the RCRA toxicity characteristic values, and 1 is lower; although it must be borne in mind that different tests are used (WET versus TCLP) so that the values are not directly comparable. The other 18 chemicals with STLCs are not on the RCRA list of chemicals with TCLP regulatory concentrations. If the concentration of any of the 36 chemicals in the waste extraction test is higher than the STLC threshold for that chemical, the waste is considered to be hazardous.

In addition to STLCs, which address leachable contaminants, the waste is analyzed to determine whether the concentration of any chemical exceeds the total threshold limit concentrations (TTLC). TTLCs are used to account for exposure pathways other than groundwater (e.g., air, surface water, and soil). For any particular chemical, the numeric value of the TTLC, expressed in milligrams per kilogram, is generally 100 times the numeric value of the STLC, expressed in milligrams per liter. TTLCs have been derived for 38 chemicals. If the concentration of any of these chemicals in a waste exceeds the TTLC for that chemical, the waste is considered to be hazardous and is subject to disposal in a class I landfill.

The current system has only one set of toxicity thresholds for STLCs or TTLCs. If a waste does not contain chemicals for which STLCs or TTLCs have been established, it may still be classified as hazardous if it is determined to be acutely toxic to mammals or fish[2] by direct testing or

---

[2]Acute toxicity in mammals is measured by dose (for oral or dermal exposure) or air concentration (for inhalation exposure) causing 50% mortality of test animals (LD_{50} or LC_{50}, respectively). The DTSC tests do not specify a duration of exposure for these definitions.

by a calculation that factors in the acute toxicity and weight proportion of its constituents. Wastes are considered to be hazardous if they have an acute oral LD_{50} of less than 2,500 mg/kg of body weight, an acute dermal LD_{50} of less than 4,300 mg/kg, an acute inhalation LC_{50} of less than 10,000 parts per million (ppm), or an aquatic LC_{50} (based on a fish bioassay) of less than 500 mg/L. An acute oral toxicity threshold of 2,500 mg of a chemical per kilogram of body weight was established by California statute on January 1, 1997 (DTSC 1998b). Acute dermal and inhalation (vapor) toxicity limits were established by multiplying the "highly hazardous" level (LD_{50} or LC_{50} values) recommended by the National Institute for Occupational Safety and Health (NIOSH 1974) by a safety factor of 100. Aquatic toxicity levels are based on the list of hazardous substances at 40 CFR 116.

A waste may also be hazardous if it contains any of 16 carcinogens at a mass fraction greater than 10 mg/kg or if data are available to suggest that a new chemical, or a new route of exposure may pose a threat to human health or the environment. The 16 carcinogens are those identified in the California Occupational Carcinogens Control Act of 1976 (California Labor Code § 9004).

Under the existing waste-classification system, DTSC regulates hazardous wastes that pose a low risk to human health or the environment in the same way it regulates wastes that pose a high risk; mitigating information on potential exposures is not factored into the classification of waste containing the chemicals. The small group of "special" wastes consists of wastes that contain only metals (on the list of 800 chemicals) that are in a non-leachable form. Again, the risks they might pose to human health or the environment are not known. Wastes that do not satisfy any of the criteria for a hazardous or special waste are not further regulated by DTSC, but may be subject to regulation by other California waste management agencies.

Need for Change in Waste-Classification System

For the past 3 years, California has been engaged in a regulatory structure update (RSU). The aim of the RSU is to reduce regulatory burdens while retaining requirements needed to protect the citizens and environment of California. The approach of the RSU is to review and update California's hazardous-waste regulations using the most current scientific data on health and environmental effects of chemicals. The RSU is based on the four guiding principles (DTSC 1997):

- Protect public health and the environment without unnecessarily hindering sustainable growth and development.
- Eliminate or modify regulations that are duplicative, ineffective, or do not provide needed protection or information.
- Foster voluntary compliance through regulatory flexibility, simplicity, and facilitating self responsibility.
- Ensure that California's hazardous-waste regulatory program reflects the current roles of other federal, state, and local regulatory programs.

To comply with the goals of the RSU, DTSC reviewed its waste-classification system, which has been in use for 12 years, and determined that changes are required. DTSC indicated that revisions to the current classification system are needed because the present TTLCs are inflexible, in that they (DTSC 1998a, p. 7):

- Require all wastes to be either rigidly regulated or completely unregulated by DTSC.
- Cannot incorporate advances in technical information.
- Provide no guidance for decisions on specific waste streams (variances or reclassification).
- Do not provide a defined mechanism for regulating additional chemicals.

Proposed Waste-Classification System

The DTSC has proposed changes to its current approach to incorporate recent advances in exposure assessment, toxicology, and risk assessment. DTSC recognizes that there is new information available on many waste constituents indicating they are more toxic or less toxic than previously thought. Moreover, new technologies have permitted the separation, reduction, and recycling of many compounds, affecting the composition of some waste and the way in which it is disposed.

According to the proposed plan, the waste-classification criteria would use a risk-based approach to create a mid-level classification for waste that is not risky enough to be classified as hazardous but too risky to be classified as nonhazardous. Thus, waste not classified as RCRA

INTRODUCTION

hazardous waste would be classified into the following three categories: (1) fully regulated hazardous wastes, (2) special wastes with reduced regulatory requirements, and (3) nonhazardous wastes with no regulatory requirements imposed by DTSC. Figure 1-1 compares the current and the proposed systems.

Under the RSU, DTSC's new regulatory approach has several goals:

- Incorporate risk into the classification of hazardous waste.
- Replace WET with the RCRA TCLP.
- Provide a step-by-step process for classifying wastes.
- Revise the criteria for special classification of waste so that they are risk-based.
- Classify waste on the basis of a constant, incremental risk threshold.

A schematic of the proposed DTSC non-RCRA waste-classification

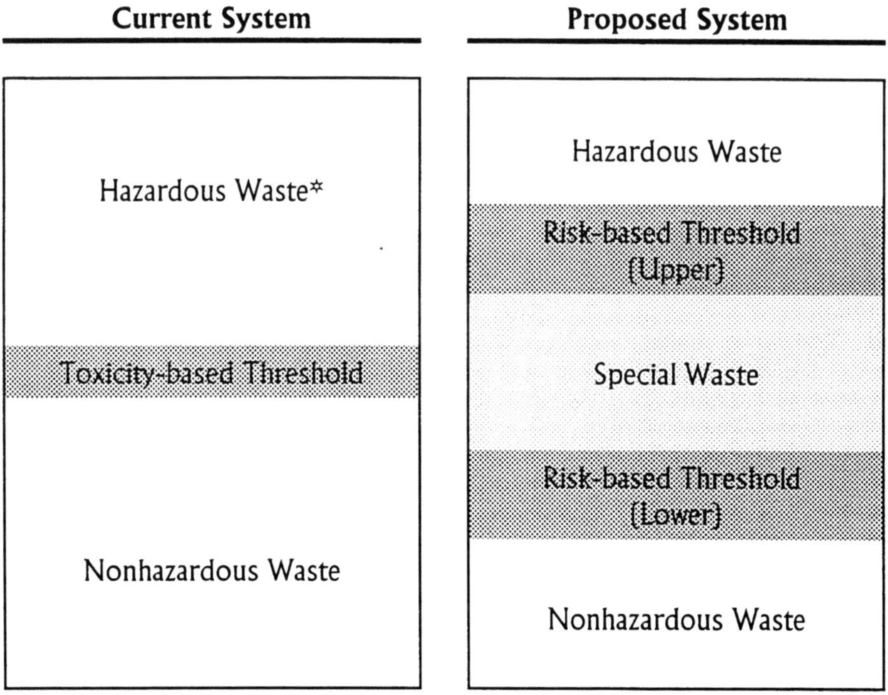

Current System	Proposed System
Hazardous Waste*	Hazardous Waste
	Risk-based Threshold (Upper)
Toxicity-based Threshold	Special Waste
	Risk-based Threshold (Lower)
Nonhazardous Waste	Nonhazardous Waste

*Under the current system, a very small number of hazardous wastes receive variances for disposal as special wastes.

FIGURE 1-1   DTSC's waste-classification systems.

system is shown in Figure 1-2. Under the proposed two-threshold classification system, DTSC would establish two risk thresholds by which wastes would be classified as hazardous, nonhazardous, or special. If the

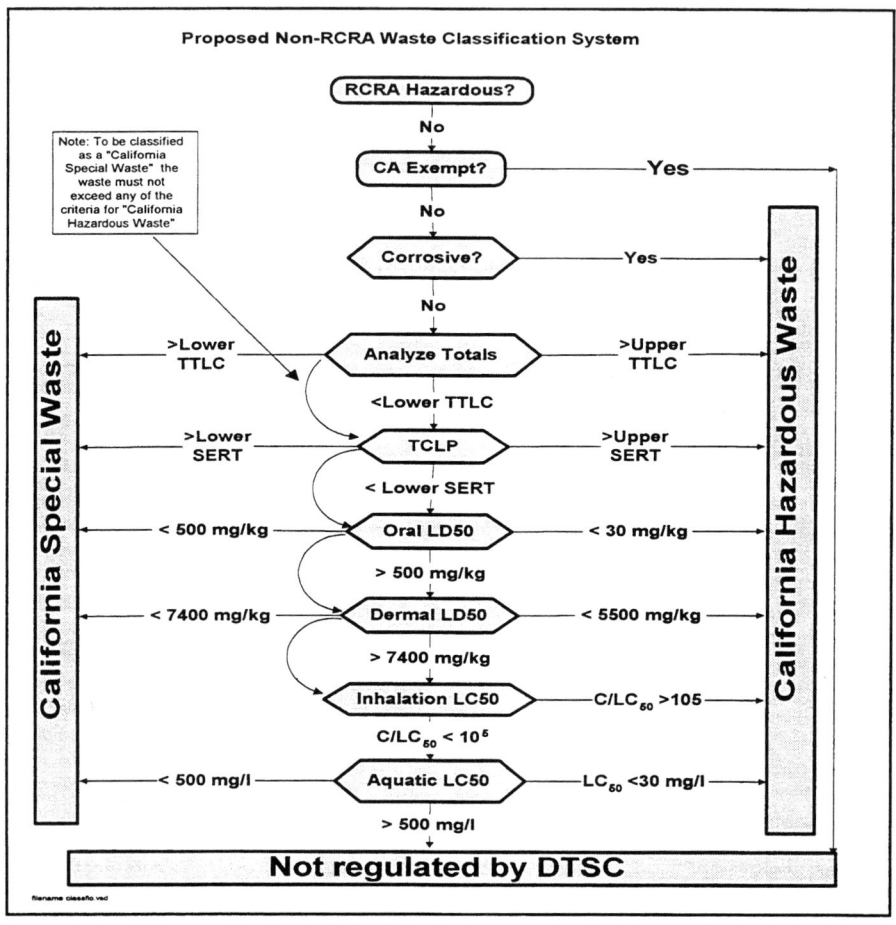

FIGURE 1-2 Proposed Non-RCRA Waste Classification System. Source: DTSC (1998a, p 36). A slightly amended version with a corrected value for the upper threshold inhalation LC_{50} was submitted to the committee by Dr. James Carlisle (personal commun., January 28, 1999). The figure omits the comparisons for toxicity of vapors and does not appear to exactly match the discussion in DTSC 1998a. Also omitted are any considerations of ignitability, reactivity, carcinogenicity, or new threats; and "CA Exempt" does not appear to be explained anywhere in DTSC, 1998a).

concentration in a waste of each regulated substance is lower than a certain threshold (an "exit" threshold), the waste would be considered nonhazardous and could be disposed to landfills other than class I landfills. If any regulated substance is present at a concentration at or above the upper threshold established for that substance, the waste would be considered hazardous and would need to be disposed to a class I landfill with a double composite liner and leachate collection system. If the concentration in a waste of all regulated substances is below the upper threshold, but the concentration of one or more of such substances exceeds the lower threshold, the waste would be classified as special. Such waste could be disposed to a landfill with a single composite liner and leachate collection system provided the waste is not determined to be hazardous by the other criteria of reactivity, corrosivity, or ignitability. Special wastes would also be subject to modified reporting and handling requirements. Risk-based thresholds have been proposed by DTSC (1998a) for the chemicals (exceptions noted) in the list shown below.

Toxicity Criteria

Total Threshold Limit Concentrations

Under the proposed DTSC scheme, the risk-based approach to TTLCs requires the identification of possible waste-management situations and the potential exposure scenarios that could result. For each exposure scenario, the possible pathways by which humans and other organisms could be exposed to chemicals in the wastes are identified. The exposure-pathway information provides input into a mathematical model that estimates the risk associated with each scenario (see Table 1-1). For the upper TTLCs (i.e., separating special and hazardous wastes), DTSC decided that an appropriate exposure scenario is disposal in a municipal solid-waste landfill comparable to a RCRA Subtitle D landfill. The lower TTLC (i.e., separating special and nonhazardous wastes) is derived to be protective of humans and ecological receptors if the waste is used as a soil amendment (i.e., the land conversion scenario).

For the upper TTLC, DTSC determined that risks are likely to be greatest for workers at the landfill and for residents living adjacent to the landfill. Risk estimates for the workers are generated by a modified preliminary endangerment assessment (PEA) model that is used to estimate exposures to both organic and inorganic waste constituents. Risks

Chemicals Proposed for Risk-Based Thresholds by DTSC	
aldrin	vinyl chloride[b]
chlordane	lead (organic and inorganic)
DDT (dichloro diphenyl trichloroethane) & metabolites	asbestos[c]
	antimony
2,4-D (dichlorophenoxyacetic acid)	arsenic
	barium
dieldrin	beryllium
endrin	cadmium
heptachlor	hexavalent chromium
kepone (chlordecone)	total chromium
lindane (gamma hexachlorocyclohexane)	cobalt
	copper
methoxychlor	fluoride
mirex	mercury
pentachlorophenol	molybdenum
polychlorinated biphenyls (PCBs)[a]	nickel
PCDD/PCDF (polychlorinated dibenzodioxins/polychlorinated dibenzofurans)	selenium
	thallium
	vanadium
toxaphene	zinc
trichloroethylene	silver[d]
2,4,5-trichlorophenoxy propionic acid (silvex)	

[a]New TTLCs have not been developed for PCBs and the current TTLC value was retained as the upper-threshold TTLC; the current STLC value was retained as the upper SERT.

[b]DTSC has proposed removing vinyl chloride from the list of carcinogens and developing both a SERT and a TTLC for this chemical.

[c]New TTLCs have not be developed for asbestos and the current TTLC value was retained as the new lower-threshold TTLC.

[d]DTSC has proposed the removal of silver from the list of TTLCs due to its low toxicity.

to residents living near the landfill are calculated using the CalTOX landfill model for organic waste constituents or the PEA model for inorganic constituents (DTSC 1998a, p. 57). The risk of chronic lead exposure, excluding carcinogenesis, for on-site workers is estimated using the DTSC lead risk-assessment spreadsheet (LeadSpread). These exposure scenarios and the models used to estimate risks are shown in Table 1-1.

For lower TTLCs, an exposure scenario is used that corresponds to residents living on land to which waste has been applied at a rate of 7,000 kilograms per hectare annually for 20 years. In this scenario, DTSC assumes that the waste has been thoroughly mixed with the upper 15 centimeters of soil. PEA, CalTOX, and the lead risk-assessment spreadsheet are used to estimate risks for this scenario.

The management of special and hazardous wastes was assumed to be adequate to protect nonhuman biota, so that only lower TTLCs needed an ecological assessment. That assessment is performed using a two-step approach. The first step consists of a comparison of the human toxicity exit concentration designated in the proposed EPA Hazardous Waste Identification Rule (HWIR) (61 Fed. Register 18780, Apr. 29, 1996) with the ecological toxicity exit concentration in that same proposal. If, for a particular chemical, the human toxicity exit concentration was lower than the ecological toxicity exit concentration in the HWIR proposal, DTSC considered that the criterion derived in the lower TTLC exposure scenario, based on human toxicity, would also protect the environment. Otherwise, a second step was used. If the SERT values for the chemicals were based on the Ambient Water Quality Criteria (AWQC) for the protection of aquatic life, then DTSC assumed again that the human-toxicity-based lower TTLC would also be protective of the environment. Otherwise, the lower TTLC was determined on a chemical-by-chemical basis.

TABLE 1-1 Models Used for the Various Exposure Scenarios

Waste Constituents	Exposure Scenario			
	Upper TTLC		Lower TTLC	
	Residents Living Near Landfill	Waste Workers	Residents on Converted Land	Ecological Concerns
Organic Chemicals	CalTOX	PEA worker	CalTOX	Multi-tiered process
Inorganic Lead	LeadSpread	LeadSpread	LeadSpread	Multi-tiered process
Inorganic Chemicals	PEA	PEA worker	PEA	Multi-tiered process

Source: Adapted from DTSC (1998a).

Once the upper or lower risk-based TTLCs are derived for any chemical using the procedures described so far, they are compared with twice the estimated quantitation level to determine if they can be adequately measured. The TTLCs are also compared with the background soil mass fractions found in California. (The background soil mass fraction is the highest mass fraction of the element found in 50 samples of native California soil (DTSC 1998a, p 51); other background values are also given in the documentation for specific chemicals (DTSC 1998a, p. 32, 52, 859).) The highest of the risk-based value, twice the quantitation limit, or the ambient background value is then considered to be the upper or lower TTLC, depending on the exposure scenario chosen.

Soluble or Extractable Regulatory Thresholds

DTSC is proposing to replace the STLCs with upper and lower soluble or extractable regulatory thresholds (SERTs) to protect groundwater. Wastes with an extractable concentration of any one or more of the designated chemicals greater than the upper SERT would be considered hazardous wastes. If the extractable concentration of each of the designated chemicals is below the lower SERT, the waste would be considered nonhazardous, providing it is not classified as hazardous by any other applicable criterion.

SERTs are based on the lowest value of the following (DTSC 1998a, p. 43):

- California maximum concentration limits (MCLs).
- EPA AWQC for the protection of aquatic life at 100 ppm total hardness and a pH of 6.5 (as applicable).
- Human-health-based levels calculated by DTSC and based only on exposure to humans from drinking groundwater.

For carcinogens, California cancer potency factors are used to derive the human-health-based levels of a chemical corresponding to a risk of 10^{-5} for an exposure scenario that appears to correspond to drinking the groundwater for the typical time that a person stays in a single residence. That risk level is the same as the risk level required by the California Safe

Drinking Water Act and Toxic Enforcement Act of 1986.[3] When California cancer potency factors are not available, the EPA cancer potency factors are used. For noncarcinogens, EPA oral reference doses or data from EPA Region 9 preliminary remediation goals are used to develop the health-effects levels (no reference for these values were provided in DTSC 1998a, p. 43). Various uncertainties, such as those involved in extrapolation from animal data to humans and the uncertainty of cancer potency factors in animals, are factored into the health-based values by using point estimates that are expected to overestimate risks (a surface-area-based interspecies extrapolation, and the 95th percentile of the experimental potency in animals).

Once the lowest value is chosen for calculation of the SERT, that number is multiplied by a dilution-attenuation factor of 100 to account for movement of the waste leachate through the unsaturated zone of soil and mixing with and dispersion in groundwater (DTSC 1998a, p. 44). The adjusted value is compared with a number that is twice the estimated quantitation limit, and the higher value becomes the exit-level SERT. In the case of arsenic, the resultant value is also compared with the 90th percentile of arsenic concentrations in drinking water supplies (which is used as a surrogate for background concentrations) and the higher value selected. For upper SERTs, the lowest value of the MCL, AWQC, or health-based level is multiplied by a dilution-attenuation factor of 100 and by a liner protection factor (to account for the reduced percolation from a landfill with a liner—the DAF of 100 was based on a landfill with no liner). Again, the resulting value is compared with twice the estimated quantitation limit, and the higher value becomes the upper SERT.

For the derivation of the SERT values, DTSC is proposing to rescind the use of the California WET for determining the extractable constituents of a waste. In lieu of the WET, DTSC is proposing the use of the TCLP that is currently used by the EPA to identify soluble toxic constituents of RCRA hazardous wastes.

---

[3]California Code of Regulations, Title 22, Division 2, Chapter 3. Safe Drinking Water and Toxic Enforcement Act of 1986. Article 6, Section 12601, Clear and Reasonable Warnings.

Other Toxicity Criteria

A two-threshold system is also proposed for evaluating acute toxicity of wastes, for human exposure by oral, dermal, or inhalation routes or for aquatic toxicity. For acute oral toxicity, wastes with an LD_{50} of greater than 500 mg/kg of body weight would be nonhazardous and wastes with an LD_{50} of less than 30 mg/kg would be hazardous; those with intermediate LD_{50} values would be special wastes. The upper threshold is based on an adult exposure scenario assuming ingestion of waste at 0.31 mg/kg of body weight. The lower threshold is based on a child's exposure scenario involving ingestion of waste at 5 mg/kg of body weight. These ingestion rates are based on the ingestion rate parameters used in the CalTOX model, with fixed uncertainty factors of 100 applied to account for extrapolation from animal data to human (factor of 10) and from lethal concentrations to minimal effect concentrations (factor of 10).

For acute dermal toxicity, the upper (hazardous) and lower (nonhazardous) thresholds are 5,500 mg/kg and 7,400 mg/kg of body weight, respectively, based on dermal LD_{50} values. As for acute oral toxicity, the upper threshold is based on adult contact with waste and the lower threshold is based on a child's exposure, with dermal contact rates derived from the dermal contact rate parameters used in the CalTOX model, and uncertainty factors of 100 applied to acute toxicity values.

For both oral and dermal toxicity classification, the LD_{50} values are either directly measured for the whole wastes, or they may be calculated from constituents of the waste.

Under the current classification system, only vapor concentrations are considered for the acute inhalation toxicity of a waste. The proposed system will account for both vapor and particulate concentrations of all chemical constituents of a waste. For volatile chemicals, the hazard classification is based on a ratio of the chemical's vapor pressure to its inhalation LC_{50}. The ratios are summed for each volatile chemical in the waste. If the sum of the ratios is less than 0.1 the waste is nonhazardous, and if the sum of ratios exceeds 1, the waste is hazardous; special wastes have sums of ratios between 0.1 and 1.

Classification of particles is more complex. DTSC proposes to classify a waste as hazardous if the respirable fraction of the waste (fraction with particle size less than 10 microns) times the sum of the

INTRODUCTION

ratios of concentrations of the individual chemicals in the respirable fraction (in milligrams per kilogram) divided by its inhalation LC_{50} value (in milligrams per cubic meter) exceeds 10^5. [Note: in the DTSC documentation (DTSC 1998a, p. 73) the upper acute inhalation value is given as 2×10^6, this discrepancy was not clarified by DTSC]. If the sum is less than 10^5, the waste will be classified as nonhazardous; no special waste category is designated for particles.

For aquatic toxicity, a lower threshold for nonhazardous wastes would be an LC_{50} greater than 500 mg/L. Hazardous wastes would be those with an aquatic LC_{50} of less than 30 mg/L. This latter value is the current aquatic toxicity threshold of 500 mg/L divided by the tenth percentile estimate of the liner protection factor (18) and rounded to one significant digit.

The current one-threshold system (at 10 mg/kg) for carcinogens would remain the same, except that vinyl chloride would be removed from the list of carcinogens and a SERT and TTLC would be developed for it.

Exposure Models

Several exposure models are used to develop the risk-based concentrations for the exposure scenarios: residents near a landfill and workers at a landfill for the upper TTLC, and residents living on converted land and ecological effects for the lower TTLCs. Descriptions of each of the computational models is given below (DTSC 1997).

CalTOX Model

The CalTOX model (DTSC 1998a, p. 105) is a multiple pathway, multimedia approach for determining the risk associated with the concentration of an organic chemical in soil. The model builds on and extends EPA's risk-assessment approaches as defined in the "Risk Assessment Guidance for Superfund" (RAGS) (EPA, 1989). The advantages of CalTOX include: (1) explicit treatment of mass conservation and chemical equilibrium; (2) calculations of gains and losses in multiple environmental media (e.g., air, soil, or groundwater) over time; (3) consideration of both transportation of the chemical between environmental media and transformation of the chemical within a medium (e.g., photolysis in air or hydrolysis in water); and (4) the capability for quantitative treatment of uncertainty and variability.

The model examines the distribution of organic chemicals in the environment and the resulting exposures of human populations via all routes and through multiple exposure pathways (including air, drinking water, food, and soil (ingestion or dermal exposure)). The CalTOX model version 2.3 is designed to model risks to humans living on or near soil with a fixed concentration of a chemical. However, this version of the CalTOX model was not considered to be appropriate for modeling risks to people living near a landfill or on residential land converted from land where there had been continuous soil amendments with waste (DTSC 1998a, p. 59).

Therefore, two modifications were made to the CalTOX model by DTSC to account for these two exposure scenarios for organic chemicals. The CalTOX landfill model relates the landfill compartment concentration to the waste concentration by allowing the use of a waste dilution factor. The model also adds an estimate of transport of the chemical in gases produced in the landfill from the landfill compartment to the air. The modifications also incorporate the differences between landfill contents and soil and the differences in areas and depths of landfills compared with residential yards. The CalTOX land-conversion model includes estimation of the root soil concentrations of a chemical from an application rate, mixing depth, application duration, and waste concentration rather than using a specific initial root soil concentration.

Preliminary Endangerment Assessment Model

DTSC stated that the CalTOX model cannot be used for inorganic chemicals as it requires estimates of the soil/water partition coefficient. The model used to evaluate inorganic chemicals and to address the landfill-worker scenario is based on DTSC's PEA model (DTSC 1998a, p. 784). Like the CalTOX model, the PEA model is an adaptation of U.S. EPA's Risk Assessment for Superfund (RAGS) methods, and therefore, the treatment of toxicity and the intake equations for the routes of exposure are nearly identical in both models. However, the algorithms used to predict the environmental fate and transport of chemicals are quite different. Whereas, CalTOX considers many exposure pathways including food, drinking water, and air, the PEA model considers only the major pathways of inhalation of vapors and dust, dermal contact with soil, and ingestion of soil. The PEA model is used to evaluate the exposure of landfill workers to both organic and inorganic chemicals, because the pathways by which a worker can be exposed to a chemical in waste are more limited and more direct than those for residents. The PEA model is also considered to be appropriate for evaluating exposure to inorganic chemicals for residents living near landfills and those living on converted land.

Lead Risk-Assessment Spreadsheet Model

Lead is evaluated differently from the other inorganic chemicals. For the derivation of SERTs, DTSC indicated that it assessed carcinogenicity using the Cal/EPA cancer potency factor, as is done for other carcinogens (DTSC 1998a, p. 775). There is no reference dose for lead, and evaluation of its chronic toxicity is based on blood lead concentrations resulting from exposure to lead in the environment. The principal difference between the lead risk-assessment spreadsheet and the PEA model is that the spreadsheet uses empirical ratios between exposure media concentrations and blood lead concentrations and adds the incremental contribution from each of the five exposure pathway to arrive at a ninetieth percentile estimate of total blood lead concentration. The latter value is combined with an estimate of a safe blood lead concentration of 10 micrograms/deciliter to back-calculate a tenth percentile soil/waste concentration, which is used for the proposed upper and lower TTLCs for lead. Adult-worker and nearby child-resident scenarios are assumed for calculating the upper TTLC, and a child-resident scenario is assumed for calculating the lower TTLC.

Potential Regulatory Impacts

DTSC has indicated that some of the changes it proposes would result in the inclusion of more wastes as hazardous, and other changes would result in the exclusion of some wastes from the hazardous category. Possible changes in the volumes of hazardous and nonhazardous waste will likely affect a broad range of regulated parties and state government agencies that deal with the regulation and management of various wastes. Municipal and county agencies might be particularly affected because the type and volume of waste disposed to some municipal solid-waste landfills might change under this proposal. DTSC has indicated that there likely will be concern about placing wastes formerly classified as hazardous in those landfills. In addition, some landfill operators might refuse to accept certain wastes whose classification changes as a result of the proposed regulations.

Changes to the method used by DTSC to classify wastes will have ramifications for other California regulatory agencies that are part of the Hazardous Waste Management Program (DTSC 1997). These agencies include the State Water Resources Control Board (SWRCB), the Integrated Waste Management Board (IWMB), and the Certified Unified Program Agencies (CUPA). SWRCB is responsible for protecting the quality of water in California by regulating the siting, operation, and closure of waste disposal sites. SWRCB oversees the discharge of wastes to land, such as application of solid waste to agricultural land. SWRCB protects water quality by classifying wastes as to their risk of contaminating the water, classifying waste-disposal sites according to their protection of receiving waters, and by adopting standards and regulations for the waste sites. The proposed DTSC method might result in some waste previously classified as non-RCRA hazardous requiring further consideration by the waste generator to determine whether the waste meets the SWRCB definition of designated waste, thus requiring disposal in class II disposal facilities.

IWMB provides for solid-waste planning and implementation and oversees local agencies that manage solid waste. Solid waste is waste not classified as hazardous. Under the proposed DTSC classification system, some wastes previously classified as non-RCRA hazardous might be reclassified as nonhazardous and, therefore, might be disposed of as solid waste to a class III landfill (DTSC 1997).

Certified Unified Program Agencies (CUPA) was initiated by California in 1996 to consolidate and coordinate the regulation and management of hazardous waste, currently mandated by six regulatory programs. CUPA is implemented by local agencies with state oversight. The six programs in the CUPA are

- Hazardous Waste Generators and Hazardous Waste Onsite Treatment Program

- Underground Storage Tanks Program
- Above Ground Tanks Program (Spill Prevention Countermeasure Control Plan)
- Hazardous Material Release Response Plans and Inventories
- Risk Management and Prevention Program
- Uniform Fire Code Hazardous Materials Management Plans and Inventories

The proposed DTSC waste-classification system will affect several of the CUPA programs and require adjustments not only by the local agencies but also by the waste generators themselves, particularly if their wastes are reclassified and subject to different management and disposal requirements.

EPA also developed a risk-based approach for the national hazardous-waste program under RCRA to identify wastes that no longer warrant being placed in the hazardous category. In 1996, EPA proposed this approach as part of its Hazardous Waste Identification Rule (HWIR) (61 Fed. Register 18780, Apr. 29, 1996). The proposed rule would have set risk-based exit levels for toxic constituents in RCRA-listed wastes generated at a facility for 192 chemicals for humans and approximately 50 chemicals for ecological receptors. The risk methodology was based on consideration of five types of waste-management units (sources); numerous release, transport, and exposure pathways; and biological-effects information. The approach to establish the exit levels was generally similar to DTSC's approach but was faulted by EPA's Science Advisory Board for several reasons, including inadequate consideration of all environmental media and setting exit levels based only on the single exposure pathway considered to be most sensitive. EPA is revising its approach, and its Office of Solid Waste has expressed interest in DTSC's approach.

2

DTSC's Proposed Overall Approach

THE NRC COMMITTEE commends the Department of Toxic Substances Control (DTSC) of the California Environmental Protection Agency (Cal/EPA) for its effort to improve the scientific structure of its regulatory system and increase the use of modern risk-analysis procedures and environmental models in this undertaking. Clearly DTSC devoted a substantial effort to the development of the proposed approach for classification of hazardous wastes, and made extensive use of many of the most up-to-date scientific procedures.

The committee based its review on the documentation and spreadsheets that DTSC provided, supplemented by DTSC's response to questions submitted by the committee, public presentations made to the committee by staff from DTSC, comments from other individuals interested in the proposed regulations, and materials available to the public on DTSC's Internet homepage [http://www.dtsc.ca.gov]. Although these materials are extensive, the committee was left with substantial uncertainty and confusion about exactly what DTSC was attempting to accomplish and the procedures and assumptions it adopted in developing the proposed regulations. Among these uncertainties are the protection goals for human health and the environment, definitive exposure scenarios (both those modeled and those unmodeled), and a logical step-by-step explanation of the actual waste classification methodology specifying how a waste is assessed based on DTSC's flowchart (DTSC 1998a, p. 36). DTSC needs to explicitly state the advantages and disadvantages of

DTSC PROPOSED APPROACH

their proposed system compared with both their current system and the federal system. The advantages in terms of regulatory flexibility, improved health and environmental protection, and any economic benefits for both the regulated community as well as the general public should be clearly articulated and documented.

In its review, the committee adopted the following criteria to guide its evaluation of the proposed methodology for classifying waste. Some of these criteria are more important to the committee's deliberations, conclusions, and recommendations than others. However, they are all important to the development and implementation of regulations designed to protect human health and the environment. The committee recommends that DTSC consider using similarly defined set of criteria when revising and evaluating their proposed approach.

- *Statement of Goals.* The effort should begin with a clear, coherent statement of the health and environmental protection goals that the waste classification system is attempting to achieve.
- *Realistic Exposure Scenarios.* The analyses should be based on exposure scenarios that are realistic and consistent with the health and environmental protection goals established.
- *Multimedia and Multipathway Assessment.* The risk assessments should incorporate, in a consistent and integrated fashion, the full range of potential exposures to human and nonhuman receptors through different media and along different pathways.
- *Valid Science.* The analyses should incorporate data, assumptions, and relationships that represent the best current scientific knowledge at the appropriate level of detail. In addition, the data and analyses should be thoroughly validated and peer-reviewed. The variability and uncertainty associated with the data, assumptions, and relationships should be explicitly recognized in the analyses.
- *Transparency.* The entire process should be clearly explained and thoroughly documented. Policy assumptions and decisions should be clearly distinguished from scientific data. The bases for making the many scientific and policy assumptions that are inevitable in such an undertaking should be clearly explained.
- *Flexibility.* The analytic tools and regulations should incorporate sufficient flexibility so that the parameters and approaches can be easily modified when it is appropriate to do so. Then situations that

differ in assumptions from the initial analyses can be realistically represented, and advances and improvements in scientific information and knowledge can be easily incorporated.

- *Implementation Practicality and Evaluation.* The proposed waste-management system needs to be designed with a continuous concern for the feasibility, practicality, and costs (to all stakeholders) of its implementation. It also requires a process for evaluating its success in achieving its goals and whether the program impacts are consistent with the assumptions that were made in its formulation.

The remainder of this chapter presents the committee's conclusions of how the approach adopted by DTSC satisfies those criteria. The remaining chapters provide more detailed comments about specific elements of the proposed classification system.

STATEMENT OF GOALS

In any undertaking directed at protecting human health and the environment, it is important to clearly set forth, at the beginning of the regulatory process, the level of protection that is being sought and for whom. These protection goals become the basic policy that drives the whole process. These goals guide the project staff on what analyses should be carried out and what types and stringency of management requirements should be imposed. They provide the information the project managers need to ensure that regulations are coherent and consistent. They also include the fundamental information the public needs to intelligently understand, review, comment on, and respond to the process and results.

The clearer and more explicit the statement of the protection goals, the more coherent the project will be. There will be less tendency for different staff members to establish their own goals, and less effort spent by the public in divining the goals of the project from its results. Clear, explicit goals should result in a more efficient development process and more focused and relevant public comments.

The statement of the protection goals should address such questions as

- What population is to be protected and what level of protection is to be provided to this population?

DTSC PROPOSED APPROACH

- What level of protection is to be provided to the most-exposed individuals?
- Which sensitive populations are to be considered and what level of protection is to be provided to them?
- What level of protection is to be provided to various forms of wildlife and other ecological components?

For illustrative purposes, the committee offers the following statements to show how the human health protection goals might be more clearly and precisely stated. For risks from hazardous waste disposal all the following protection goals would be required:

- Ninety-five percent of California's population will be protected from carcinogenic risks to a level of 10^{-6} or better (lifetime risk).
- The reasonably most-exposed children will be protected from carcinogenic risks to a level of 10^{-5} or better (lifetime risk).
- All other reasonably most-exposed individuals will be protected from carcinogenic risks to a level of 10^{-4} or better (lifetime risk).
- No reasonably most-exposed individual will be subject to other chronic health risks at a level greater than a hazard index[1] of 1.0.
- No reasonably most-exposed individual will be subject to acute health risks exceeding a margin of exposure of 10.

"Reasonably" might be defined as assuming that existing regulatory programs are implemented effectively, exposure assumptions are consistent with existing identified conditions, and populations and individuals

---

[1] Hazard index is generally considered to be the quantity derived to estimate the likelihood of adverse effects from exposure to noncarcinogens. It is calculated from the following equation:

$$HI = \frac{\text{Average lifetime dose (mg/kg/day)}}{\text{Reference Dose (mg/kg/day)}}$$

A reference dose is considered to be the level of a chemical that will not lead to adverse health effects over a lifetime of exposure. A hazard index of greater than 1 indicates a potential risk. Hazard indices may be added for multiple chemicals if they have the same adverse effect.

are considered to have no unusual sensitivities to the substances to which they are exposed.

The last suggested statement of a health protection goal concerns acute health risks. DTSC's inclusion of acute toxicity measures in its classification scheme was confusing, as no goals involving acute effects were given in the DTSC documentation. It appears, however, that based on the proposed classification schematic (DTSC 1998a, p. 36) DTSC does have a distinct goal of protecting against acute effects, in addition to the partially stated goal of protecting against chronic effects. The committee recommends that DTSC establish protection goals at some level for both acute and chronic effects.

Similarly, for wildlife (flora and fauna) protection, the goals might be stated as protecting 95% of the individuals in 90% of the species by not exceeding a critical concentration 90% of the time (or over 90% of the state's area). To make such a goal operational, DTSC would probably have to identify, for example, by species sensitivity and exposure assessments, certain "indicator species" that would represent the entire exposed wildlife population.

California law requires DTSC to provide protection from "substantial present or potential hazard" even when the wastes are "improperly treated, stored, transported, or disposed, or otherwise improperly managed" (California Health and Safety Code §25141 (a)(2)). Such a prescription might be built into DTSC's goals by stating the allowable levels of risk to individuals, under the assumption that other controls are not implemented effectively.

The documentation presented to the committee does not contain a clear exposition of goals for either the human health or ecological protection. The document refers to a "reference carcinogenicity risk level" of 10^{-5} (DTSC 1998a, p. 3), and, in an appendix, states that "all health-based values are 10th percentile estimates of the concentrations that would correspond to the stated level of risk or hazard" (DTSC 1998a, p. 43). In oral and written responses to questions from the committee, DTSC indicated that "the intent is to be 90% certain that the true risk or hazard index for an individual taken at random from the population is not greater than 10^{-5} or one, respectively." (DTSC, personal commun., October 29, 1998, see Appendix C, No. 28, question 71). This is equivalent to setting a goal of protecting 90% of the population of California from risks equal to or greater than those indicated.

However, most of the critical-exposure estimates (i.e., those used for determining regulatory levels) appear to be based on the most-exposed individuals. Thus, it is unlikely that 10% of California's population consists of landfill operators with 30% of their skin exposed, or that 10% of the population lives within 100 meters (m) of a landfill (assumptions made in the exposure scenarios). The analyses apply the 10^{-5} carcinogenic risk protection and the hazard index of 1.0 to these most-exposed individuals without explicitly enunciating any health protection goals for them. It is also unclear whether the individual maximum exposures used are reasonable for the chronic exposure scenarios, and there are no clear scenarios propounded for acute exposures.

With respect to wildlife, the analyses fail to establish clear wildlife protection goals, and they appear to assume in many cases that risk levels established to protect human health will also protect wildlife. DTSC provides no justification for this assumption.

The committee notes that definitive data are rarely available to answer risk-related questions unambiguously. As a result, many risk-based decisions are made on the bases of science-policy decisions. These science-policy decisions can include, but are not limited the selection of protection goals, exposure scenarios, sensitive subpopulations, and default assumptions. While evaluation of policy is not within the committee's task, such decisions must be recognized as forming one foundation of DTSC's proposal. They cannot be ignored when reviewing the scientific merits of the proposal.

MULTIMEDIA AND MULTIPATHWAY RISK ASSESSMENT

The existing DTSC waste-classification system is broader than the federal waste-classification system under the Resource Conservation and Recovery Act (RCRA), in that the DTSC system considers acute toxicity, effects on aquatic life, and human exposure pathways other than groundwater. Although the current waste-classification system is broader than RCRA, it is based only on the potential toxicity of the waste and its constituents, and does not factor in potential exposure, including environmental fate and transport. The waste-classification system being proposed by DTSC attempts to correct those deficiencies, making use of multimedia, multipathway, risk-assessment models to classify wastes.

DTSC's proposed approach differs from the current federal RCRA waste classification in three ways. First, the proposed system has two thresholds for the classification of wastes, rather than a single threshold. Second, the proposed approach is based on risk, rather than solely on the toxicity of the chemical of interest. Thus, both the toxicity of the chemical and the potential for exposure are considered in the proposed waste-classification system through the use of multimedia, multipathway risk-assessment models. Third, the proposed system is based on individual chemicals rather than on sources or types of waste. RCRA has a list of approximately 40 chemicals that are regulated on a specific-chemical basis, but the rest of the classifications are based on sources or types of waste.

DTSC's efforts to improve the state's environmental management by using a more comprehensive, multimedia, multipathway, risk-based approach is commendable. However, no waste-classification system can be entirely risk-based. DTSC should reconsider two of the decisions it made in the selection and use of risk-assessment models. The CalTOX model incorporates many of the appropriate factors for undertaking a good multimedia, multipathway risk assessment, and its components can be considered to approximate the state-of-the-art when correctly applied. A disadvantage of CalTOX is that it does not include exposure scenarios that are explicitly appropriate for ecological receptors. Given the overall advantages of such a model, however, the committee questions why it was not used more extensively. In the analyses, exposures to water-soluble contaminants and non-water-soluble contaminants are evaluated separately in soluble or extractable regulatory thresholds (SERTs) for groundwater pathways and total threshold limit concentrations (TTLCs) for air, surface water, and soil pathways. However, human and ecological receptors are typically exposed to contaminants by more than one pathway. It is desirable, therefore, to consider multimedia, multipathway exposures using a single model rather than to evaluate different pathways separately; DTSC should consider integrating the exposure calculations to simultaneously include all pathways and environmental media. For example, benzene is a constituent of many wastes and is readily mobile within and between most environmental compartments. To determine the risks associated with benzene contamination of a waste, its movement through and between air, soil, groundwater and surface waters should be assessed simultaneously.

DTSC has decided to classify wastes on the basis of single chemical components. The approach has a clear practical advantage in that information on the toxicity of specific chemicals is generally available, whereas toxicity information about wastes from various sources is generally unavailable. Also, the chemical composition of waste from specific sources or processes can vary from time to time both qualitatively and quantitatively. Classifying waste on the basis of individual chemicals allows the use of available toxicity data, response to changes in waste composition, and addition of new information.

However, the committee questions DTSC's single-chemical approach. The disadvantage of regulating on a chemical-by-chemical basis is that people and ecological receptors are rarely exposed to individual chemicals; instead they are typically exposed to mixtures of chemicals. The proposed classification system does not address the additive or potential synergistic effects of exposure to multiple toxicants. The U.S. Environmental Protection Agency (EPA) handles the risk of mixtures by adding the cancer risk estimates from each component and, for noncancer risk, by adding the hazard indices for each component, at least for a first approximation. DTSC has chosen not to add the risk estimates from each chemical in a mixture when defining TTLCs or SERTs. Such an approach may underestimate the risk from a mixture. It is recommended that DTSC, lacking better information about the effects of chemical mixtures, adopt the same default assumption as EPA, and linearly add the risk estimates for the mixture's components to estimate the risks associated with the entire mixture. DTSC is encouraged to monitor any scientific advances in the evaluation of chemical interactions at low exposures and to incorporate such advances into their approach. DTSC's proposed approach does, however, consider whole wastes, or all waste components, when assessing risks from acute toxicity (see Chapter 4, Toxicity Tests).

Federal regulatory decisions are typically a blending of risk-based procedures together with a multitude of policy decisions. DTSC's approach is no different. For example, the structure of the risk-assessment model to be used (in this case a model of the migration of a chemical between environmental compartments), the particular parameters to be used in the model, and the exposure scenarios to be employed all involve policy decisions by DTSC. A completely risk-based approach to regulations is impractical, if not impossible. Chapter 4 of this report contains substantial detailed comments on DTSC's selection and application of the risk-assessment models.

REALISTIC EXPOSURE SCENARIOS

The problem of inadequate explication and justification of policy assumptions incorporated in DTSC's analyses was particularly noticeable in the selection of the exposure scenarios. The selection of realistic exposure scenarios is fundamental to carrying out reasonable risk assessments. However, the committee had great difficulty in examining the appropriateness of the scenarios selected by DTSC, because there was no discussion of the policies that DTSC wished to implement, or documentation of any connection between policy and scenario. The DTSC staff, in written and oral responses (DTSC, personal commun., October 9, 1998, see Appendix C, No. 28; DTSC, personal commun., November 20, 1998), suggested that there had been some process by which the scenarios were selected; however, they provided no information on what the selection process was and how or why some scenarios were rejected and others were not. DTSC also suggested that the scenarios were selected to be representative not only of the explicitly described situation, but also representative of other situations (e.g., the landfill scenario was also supposed to apply to waste piles). However, no documentation was provided to elucidate the other situations that each scenario was supposed to represent. The description of the acute exposure scenarios was extremely limited for oral and dermal exposures and non-existent for inhalation exposures, and the committee is unsure whether DTSC intended to protect separately against acute effects, or whether the acute toxicity part of the classification scheme was somehow also supposed to protect against chronic effects.

The justification of the wildlife exposure scenarios was even weaker than that of the human exposure scenarios. No wildlife-specific scenarios were provided by the DTSC, and it seems likely that human exposure scenarios are inappropriate for wildlife. The DTSC does not provide a basis for their assumption that the lower TTLCs or SERTs derived for the protection of humans would be protective of aquatic or terrestrial wildlife. Although there are some differences in sensitivities between species, the greatest difference in toxic effects of a chemical on wildlife often results from differences in exposure of wildlife. Some wildlife, because of restricted home ranges, are likely to have higher exposure than humans. This is particularly important for soil-dwelling organisms or those at the top of food chains where the biomagnification potential of chemicals could come into play.

The DTSC report does not provide any details about how ecological assessments will be conducted. The report does indicate the general steps that would be taken for conducting a de novo ecological risk assessment, but it does not specify the scenarios or the methods that would be used to derive reference doses for specific chemicals to compare with estimated environmental or tissue concentrations (DTSC 1998a). Chapter 3 of this report contains other observations about DTSC's selection of realistic exposure scenarios.

VALID SCIENCE

The committee is pleased that DTSC has incorporated a number of up-to-date and valid scientific methods in developing its proposed regulations. These include not only the multimedia, multipathway risk assessments mentioned earlier, but also sophisticated modeling techniques and a statistical method for incorporating factors whose values vary for different members of a population. In making these selections, DTSC has established an overall scientific approach to its regulatory effort which, in many ways, represents the best current scientific knowledge. However, in many of the details of its effort, DTSC has failed to realize the potential promised by this approach. These failures are the focus of Chapters 3 and 4. In general, DTSC's approach would have benefitted substantially from a careful internal review for content, presentation, and implementation prior to external peer review.

The information presented to the committee regarding the scientific analyses is very uneven, and the poor documentation has made it difficult to judge the validity of the scientific methodologies adopted. The documentation for the principal exposure model used in the analysis, CalTOX, appears to have an adequate level of detail, although the committee has not performed an extensive peer review of all model details. The computations in the lead spreadsheet are adequately documented, but the tables are confused by the inclusion of extraneous materials, such as the inclusion of non-zero values for adults when the receptors selected by the scenario description are children. In addition, the structure of the model might not be clear to the uninitiated reader. The documentation of the preliminary endangerment assessment (PEA) model is inadequate and contains many errors. It is difficult to ascertain what values were used for which scenarios, and there are substantial differences between the documentation and the spreadsheet implementation. The documentation for calculation of the SERTs is adequate, although it is impossible to determine what distribution was supposed to be used for the liner protection factor.

The documentation for the acute exposure scenarios is so limited that the committee was unable to discern DTSC's reasoning, and hence was unable to tell whether those scenarios were appropriate or appropriately used. If the acute scenarios were indeed designed with a goal of protecting against acute risks (separately and independently of the goal of protecting against chronic risks), then the approach taken might be reasonable. Even then, use of an endpoint such as an acute lowest observed adverse effect level would be preferable to extrapolating from an acute LD_{50} or LC_{50} value. If, however, DTSC is attempting to use the acute toxicity measures for protection against chronic risks (and the acute toxicity approaches are currently all that are applied for any wastes that do not contain any of the 38 TTLC chemicals), then the measures of toxicity used by DTSC are scientifically inappropriate.

DTSC's efforts to adopt a statistical method for characterizing variables is to be commended. However, the application of this method was incomplete and uneven. Many factors, particularly those characterizing exposure scenarios and other policy decisions, which could have been characterized as statistical distributions, were, instead, included as point estimates. DTSC also failed to address adequately the distinction between variability and uncertainty. It was stated that the 90th percentile of the calculated distributions was to be used; however, there was no distinguishing variability and uncertainty distributions in the discussion or in the analyses. Such analyses may have meaning only if the calculations are for randomly selected individuals (so that the variability becomes an uncertainty). In response to the committee's questions, however, DTSC indicated that they were aiming at the 90th percentile of both the variability and the uncertainty distribution; this is not achieved using the statistical method they selected.

Another major failure, discussed in more detail in Chapters 3 and 4, is the inadequacy of the sensitivity analyses. DTSC's proposed approach is complex, and involves many assumptions. Sensitivity analyses are essential to determine the most important parameters of the models, and afford a mechanism to focus on those important parameters. During the presentations to the committee, comments on several of the more unreasonable scenarios presented in public comments were dismissed by

DTSC PROPOSED APPROACH

DTSC by allusion to sensitivity analyses that indicated a single or small number of pathways dominated the risk. However, the presentation of such sensitivity analyses was weak and the analyses themselves were insufficient to justify the exclusion of some pathways. It is important that DTSC determine how sensitive the results of the analyses are to the numerous factors and assumptions in the model. DTSC has failed to do this in a comprehensive manner, and has failed to present and interpret the sensitivity analyses it did undertake in a meaningful and readily understood manner.

The documentation of the values used in DTSC's analyses is uneven. In some, there is substantial explanation of the values used; in others, very little. Sometimes a value is selected based on a reference to its use elsewhere, although the referenced document does not provide any indication of the source of the specified value (e.g., air-dispersion mixing height above a landfill, where the value selected might be reasonable as used in the referenced document but is incorrect in the DTSC context).

In its attempt to use the most current scientific approaches in its analyses, DTSC made use of models, data, and scientific assumptions that have not been adequately validated or peer-reviewed. In one such case, DTSC used methodologies and data from EPA's proposed hazardous waste identification rule that were found to be inadequate by EPA's Scientific Advisory Board when subjected to peer review. If DTSC believes that it needs to make use of information or methodologies that have not yet undergone full peer review, it should thoroughly document these cases and, to the extent possible, subject them to peer review before releasing the proposal for public comment.

DTSC also should have included the following information in its documentation:

- An analysis of the physical processes involved in the transport of waste from source to receptor in each scenario, and the models that are to be used to represent such physical processes (model requirements).
- A complete (mathematical) description of the models that are used; and a specification of which models are used in which scenarios to match which physical processes.
- A complete specification of each parameter value used in the document, together with references allowing retrieval of the description of its measurement or its method of calculation.

TRANSPARENCY

DTSC has undertaken a very sophisticated and complicated analysis in support of its proposed methodology. This creates a burden on DTSC to clearly explain the complex models and risk-assessment processes used to establish its regulatory limits. Most of the community affected by these regulations has little capacity (including time and training) to fully understand the basis for or implications of the proposed regulations. After reviewing almost 2,000 pages of materials (including public comments), the committee concludes that DTSC has failed to provide a transparent explanation or justification of the proposed methods, including a logical step-by-step explanation of how a waste is classified (DTSC 1998a, p. 36).

The documentation provided is incomplete, disjointed, overly long, contradictory in places, and confusing. It appears to be predominately an assemblage of disparate documents, at least some of which were still only in draft form. Although DTSC claimed to be under a tight schedule, this schedule was largely self-imposed. The number of years that it has taken to prepare this proposed approach should have provided an opportunity to prepare adequate documentation. Preparing documentation should not be postponed until the last minute; it should accompany the release of the proposal for public comment. Evidence of the inadequacy of DTSC's explanation of its approach lies in the fact that many of the public comments provided to the committee appear to be based on incorrect interpretations about what DTSC is proposing and why. The principal documentation failures, apart from missing elements, are the lack of logical order to the documents, the lack of a complete overview with adequate cross-referencing, and the lack of adequate copyediting for accuracy and clarity. The absence of table of contents, index, and pagination, incorrect cross-referencing, and inadequate referencing to source documents are also indicative of poor documentation.

Assuming that the selected scenarios match the policy requirements, a link between the models and the scenarios in which they were used, including a clear and complete description of each scenario, is lacking. Although in some cases, it is clear which models were used in which scenarios, there is a failure to match model requirements with the model used (e.g., dust emission and dispersion, landfill organic emissions, liner protection factor, and selection of many parameter values, such as dust-deposition rate, and mixing height above landfill). This indicates that

DTSC PROPOSED APPROACH

DTSC's process for matching model requirements to scenarios was flawed. It appears that DTSC used existing models that were close to what was required for the scenarios, perhaps after modifying them slightly. No analysis of model requirements for each scenario is documented, however, so the committee can only note where the models used do not match the scenarios selected.

The selection of values used in the models is often not adequately described or referenced; and sometimes the values used do not correspond to the descriptions or references cited (e.g., the values for the mixing depth of soil, and the deposition rate of dust). In some cases, the mismatch occurs because the scenarios were not adequately described. For example, the potential dust-emission area for the landfill was set equal to the whole landfill area. DTSC indicated at the second public meeting (DTSC, personal commun., November 20, 1998), that this value was selected because of an assumption that special waste could be used as landfill cover; or perhaps because this scenario was also supposed to account for the waste pile. There is no documentation of either possibility; such documentation belongs in the missing discussion of the relationship between policy and scenarios.

The most serious documentation failures relate to the policy decisions that are necessarily a fundamental element of these analyses. These flaws range from what level of health and environmental protection is intended, to why twice the level of detection was established as the regulatory level when the risk level falls below the detection-level criterion. The committee is still unsure about the DTSC's intended approach to background concentrations because different approaches appear to be suggested at different points in the documentation, and the practice appears to be different from any of them.

It is recommended that the documentation for the proposed regulations be substantially improved to clearly set forth:

- An explanation and justification for the policy decisions incorporated into the proposal.
- An explanation of the regulatory context or scientific basis for these decisions.
- A description of the process used to relate the policy decisions to the scenarios to be examined. In particular, how are all exposure scenarios reduced to a finite set of exposure scenarios (which may be a set that represent the worst case in some sense)?

- A complete description of all the selected scenarios to be examined, including the assumptions related to any policy decision.
- Several examples of known wastes that illustrate the differences between the current and the proposed classification systems.

In addition, it is strongly recommended that the documentation include a table of contents, consistent pagination and cross-referencing, perhaps an index, and an introduction that summarizes the DTSC waste classification process in toto, and the layout of the documentation, with cross-references.

FLEXIBILITY

DTSC's proposal intends to establish a waste classification system that is more flexible than the one currently used. For example, the use of three classifications of wastes instead of two provides additional flexibility in how wastes are handled. It is not economically reasonable to treat all hazardous wastes as if each one posed the same risk to humans or the environment, and even among the hazardous wastes there are degrees of hazard and risk. Different requirements for the disposal of the three waste types should allow flexibility in the stringency of the waste-disposal requirements. However, in practice, the one "bright line" currently used to separate hazardous from nonhazardous waste is usually a "fuzzy line" due to the uncertainties associated with the calculation of the value for the line. Therefore, it is not clear if one could truly distinguish two separate values for classification of wastes without an overlap of the confidence intervals around the lines.

However, inadequate consideration has been given to the need for flexibility in design and implementation of the regulations, given the high variability in the conditions to which the regulations are to apply, and the changes that are likely to occur in the scientific underpinnings of the regulations. Indeed, in some instances the proposal appears to make it more difficult to respond to such changes. DTSC needs to clearly articulate the advantages and disadvantages of the proposed system and to demonstrate where the proposed system provides greater scientific and regulatory flexibility and who will be affected by it.

For example, the proposal appears to lock in place a list of 38 chemicals of concern that DTSC apparently created some years ago. No criteria for selection of these chemicals are given in the DTSC documentation (DTSC 1998a); however, during the public meetings, DTSC did indicate that these chemicals were selected solely for historical reasons. The list of chemicals for which TTLCs and SERTs are being proposed is very constricted (see Chapter 1). Although economically important at one time, the production and use of many of the organic chemicals (most of which are pesticides, such as DDT) have ceased for several years. Therefore, although still environmentally relevant because of their persistence in contaminated materials, such as soils and sediments, they are unlikely to be present at significant concentrations in current industrial wastes. The list of TTLCs and SERTs does not include values for other potentially hazardous materials, such as the xylenes or phthalates, that could and do commonly occur in wastes.

Although one of DTSC's goals is to regulate hazardous chemicals other than those on the TTLC and SERT lists, there is currently no proposed methodology to add additional chemicals to those lists. The documents submitted to the committee refer to a list of 800 chemicals, but there is no indication of whether the risks of these 800 chemicals would be further evaluated, nor of how such an evaluation would be conducted. The application of a limited list of, at most, 38 TTLCs or SERTs to the classification of wastes is a significant flaw in the proposed waste-classification system. Application of the proposed waste-classification system to a much larger list of chemicals would probably require significant modification of the classification methodology to properly account for the wide range of properties of those chemicals.

DTSC should establish a well-defined process for adding substances on its list of regulated chemicals, and for incorporating advances in scientific understanding regarding the risks that these substances might pose. Perhaps DTSC can begin the process of including additional substances before issuing the final regulations because making frequent changes in regulations such as these is expensive and is likely to leave the public confused about what is being regulated and what is not.

The committee is also concerned about how flexible the proposed regulations will be in dealing with special wastes and the particular characteristics of specific waste-management proposals. DTSC has indicated that it intends to provide such flexibility through a variance process. However, variance processes are rarely as responsive, quick, and efficient as anticipated. This is particularly true when no special provisions have been made in the regulations to establish a responsive process for reviewing and issuing variances. In the absence of such a process, variances are likely to be subject to an often burdensome management process. The relatively few variances that DTSC has apparently issued in the past reinforces this concern about the lack of such flexibility.

The reliance on a variance process appears to introduce another weakness into DTSC's proposal. Variances are usually requested by waste generators or managers when they believe that the characteristics of their waste or its proposed management warrant a relaxation from the stringent regulatory requirements. However, these parties have no incentive to request a variance when the characteristics of the waste or its proposed management warrant a tightening of the requirements. Individuals that might have such an incentive, for instance families living adjacent to a landfill, lack the information and resources to request a variance. Thus, an efficient variance process might adequately respond to situations in which the regulation imposes excessive controls (e.g., reclassification of a hazardous to a special waste) but not in which it provides inadequate controls (e.g., reclassification of a non-hazardous waste to a special waste).

IMPLEMENTATION PRACTICALITY AND EVALUATION

Ultimately, the most important concern regarding the quality of a waste-management system is whether it provides adequate protection to human health and the environment? Other important concerns include: how effective and efficient is its implementation? does its implementation impose excessive delays and other costs on the public? can it be easily understood and followed by those who have to comply with it? and can it be easily enforced? The answers to such questions cannot, of course, be fully known until the proposed regulations have been adopted and implemented. However, fundamental questions such as these should guide the regulatory development.

The committee considered these questions and asked DTSC staff to address them at the public meetings. The DTSC answers were not always clear. To a large extent this is because DTSC apparently distinguished between the current regulations that classify wastes and any future waste-management regulations issued pursuant to this proposed waste-classification scheme. Based on this distinction, the implementa-

DTSC PROPOSED APPROACH

tion questions relating to the waste-classification regulations of the program are less important than those relating to the subsequent waste-management regulations. The primary implementation questions regarding the waste-classification procedures are the following:

- Can waste generators and managers clearly understand how they are to classify their wastes (e.g., what tests have to be conducted)?
- Are these waste-classification procedures likely to provide accurate classifications?
- Are these the least costly and most efficient procedures for obtaining the necessary information at the desired degree of accuracy?
- Can compliance with the regulation be quickly and easily determined?

No information has been provided to the committee that would allow it to make judgments on any of the above questions. However, the committee notes the following:

- The distinction between the waste classification regulations and management regulations is apparently not complete. The exposure scenarios necessarily incorporate assumptions on how the wastes will be managed, and thus presume the results of the waste management regulations. Indeed, according to an explanation given in response to a question (DTSC, personal commun., October 9, 1998, see Appendix C, No. 28), DTSC presumes that all hazardous wastes will be disposed to a class I landfill, and all special wastes will be disposed to a class I landfill, or a class II landfill if DTSC issues a variance. Thus, it appears that these classification regulations actually incorporate the waste-management decisions, although no pertinent information about the proposed waste-management regulations was provided by DTSC.
- The analyses and scenarios are limited to the disposal of solid wastes. But the definition of solid wastes, at least in federal law, also includes many liquids. This is also the case in California because the proposed classification system explicitly includes sewage sludge, which has a high liquid content. It is not clear whether the analyses on which the proposed regulations are based are appropriate for liquid wastes. It is clear, however, that the landfill disposal assumptions used to determine the special waste and hazardous waste thresholds are not. Under federal law, most liquid wastes containing hazardous components cannot be disposed to landfills.

- It is not reasonable to assume that the volume of waste classified as non-RCRA hazardous waste will not change significantly under the proposed regulations. Enough questions have been raised by public commenters to call into question DTSC's projected estimates of current versus future regulated-waste volumes (DTSC, personal commun., October 9, 1998, see Appendix C, No. 28, question 15 and attachment 3).

DTSC should give further consideration to the following factors in implementing the regulations:

- Establish a public process for evaluating the effectiveness and efficiency of the regulations. Are the regulations providing the desired degree of protection at the least possible cost?
- Establish incentives and processes within the regulatory scheme for improving the information on which the regulations are based. For example, some of the regulatory levels have been determined by current limits of analytical detection rather than human health or environmental risks. It would be advantageous to create, within the regulatory framework, incentives for improving analytical procedures so that the regulations can become more truly risk based. Without such consideration, the regulations might create strong incentives against such analytical improvements.
- Establish, as discussed above, efficient processes for incorporating new information into the regulations and for allowing the regulations to be tailored to site- and waste-specific characteristics. This is particularly important if the regulations do, in fact, incorporate waste management requirements and are not solely a waste-classification scheme.

3

Scenario Selection and Modeling

IN THIS CHAPTER, the NRC committee summarizes the general purpose of exposure scenarios, the reasons for using environmental transport models, and the specific scenarios and models employed by the Department of Toxic Substances Control (DTSC) of the California Environmental Protection Agency (Cal/EPA). A framework for evaluation of the exposure scenarios and aspects of the models that DTSC used are discussed. The focus of the chapter is the general processes involved in scenario selection and development of models, rather than particular details. Evaluations of some specific aspects of the DTSC scenarios and models can be found in Chapter 4.

EXPOSURE SCENARIOS: PURPOSE

People, animals, plants, and lower organisms (collectively, receptors) are, or can be, exposed to contaminants originating from waste materials in a variety of ways and to varying degrees. Moreover, exposure pathways will vary with time, as wastes are handled differently, and as the behaviors of the exposed populations change. Predicting every way in which any receptor might be exposed is a never-ending and ever-changing task. Despite this, the fundamental aspects of exposures are often fairly constant; and the exposures of a large number, sometimes the great majority, of different receptors can be represented with fair accuracy by abstract exposure scenarios that are representations incorporating the most important of those fundamental aspects.

The purpose of exposure scenarios is to provide abstract representations of real-world exposure situations in sufficient detail and in sufficient number to capture all the important exposure pathways. As emphasized in Chapter 2, the risk-management goals drive the selection of the exposure scenarios; and an adequate selection of scenarios is essential for the validity of a risk assessment. Even if the best possible models of environmental fate and transport were to be used, a poor selection of exposure scenarios might invalidate any conclusions drawn from the assessment. Poorly determined scenarios might result in risk estimates that are not remotely related to situations likely to occur in the real world, or in the evaluation of risks to the wrong set of receptors. For DTSC's purposes, assumptions regarding proximity to landfills, types of likely exposures (e.g., contact with contaminated soil, or through inhalation of airborne contaminants), and duration of exposure are critical to developing an accurate risk profile.

MODELING: PURPOSE

The use of environmental fate and transport models to predict environmental concentrations is required due to a lack of comprehensive measurements of contaminants in situations of interest as well as the fact that the risk assessments are often conducted a priori. Available measurements (either from the field or from laboratory experiments) are used to extrapolate to probable situations. In DTSC's risk-based approach to classifying wastes, the modeling is required to estimate environmental concentrations of contaminants in certain well-defined exposure situations (the exposure scenarios). The models are used to evaluate the extent to which receptors (human, for the health-effects modeling; animal, plant, or lower organisms, for the ecological-effects modeling) would be exposed to chemicals contained in the waste. The object is to choose models that account sufficiently accurately for the physical processes that occur in transport of the contaminants along the major pathways of exposure.

SUMMARY OF DTSC EXPOSURE SCENARIOS

DTSC proposed four exposure scenarios to model risks of wastes not covered under the federal Resource Conservation and Recovery Act (RCRA). A summary of these scenarios, based upon the DTSC report (1998a, pp. 56 ff.), follows. Given the concerns about DTSC's proposed approach as discussed in this report, the committee does not endorse the accuracy, completeness, or appropriateness of these scenarios or the following descriptions, which are based upon the DTSC documentation. Deficiencies in the scenarios and their descriptions, and instances where DTSC has varied from its stated intent, are indicated below.

Adjacent Resident Scenario

Residents are assumed to live 100 meters (m) from the fence-line of an operating landfill accepting wastes. The waste is assumed to contain up to 100% of the specific chemical contaminant under analysis. Landfill size and soil and waste conditions are consistent with those found at California landfill sites. Various transport mechanisms for the chemical contaminants are included in the modeling. The chemicals might be transported through the air as a vapor or attached to dust particles. The vapor might be inhaled, and the particles might be inhaled or deposited on backyard soil and vegetation. Residents might come into direct contact with resulting contaminated soil and they might consume foods contaminated by chemicals taken up from the contaminated soil or deposited dust particles. For some chemicals, transfer from a mother's blood into breast milk for ingestion by nursing infants is also taken into account.

Waste Worker Scenario

Workers handling the waste are assumed to handle undiluted waste directly, either before arrival at the landfill or at the landfill and without (as far as the committee could determine) any assumed institutional health and safety controls. Pathways of exposure are: inhalation of dusts or vapors; inadvertent ingestion of the waste; and dermal contact with the waste, leading to dermal absorption.

Land Conversion Scenario

This scenario replicates the assumptions of land application used by the

U.S. Environmental Protection Agency (EPA) in deriving maximum concentrations of contaminants allowed in biosolids (40 CFR Part 503). The assumptions include an application of waste to the land surface at a rate of 0.7 kilograms of waste (kg) per square meter per year for 20 years. After the last application, residents are assumed to occupy homes built on the land. The residents have attributes and exposure pathways similar to those in the adjacent resident scenario, but are additionally assumed to eat fish from a surface water body adjacent to the land on which the material was applied.

Ecological Effects Scenario

Ecological effects were addressed using a 2-step approach. The first step relies on the ecological toxicity exit concentrations developed for the proposed EPA Hazardous Waste Identification Rule (HWIR) (60 Fed. Regist. 66344, Dec. 21, 1995). HWIR is an EPA program similar to DTSC's program to categorize waste according to risks arising from its disposal. In the HWIR, EPA proposed waste concentrations for chemicals based on residential exposures as well as ecological effects. Many of the chemicals for which DTSC has proposed total threshold limit concentrations (TTLCs) are included in the HWIR. For those chemicals common to both HWIR and DTSC, if the concentrations proposed for residential protection in the HWIR proposal are less than those based on ecological effects in that same proposal, no further action would be necessary by DTSC. For such chemicals, the residential protection concentrations in the DTSC proposal would be considered to be protective for ecological effects. As a second step, for the remainder of the chemicals with ecologically based HWIR-proposed values, DTSC has developed ecologically based TTLCs.

MODELING USED IN THE SCENARIOS

The modeling efforts used in these four scenarios are summarized in Table 3-1. One of those efforts was a modification by DTSC of CalTOX, a multipathway, multimedia model (DTSC 1998a, pp. 78-647). It is implemented through a set of Microsoft Excel spreadsheets, using data summarized specifically for CalTOX (DTSC 1998a, pp. 182, 555, 591),

TABLE 3-1 Models and Spreadsheets Used by DTSC for the Scenario-Chemical Combinations[a]

Criteria	Upper TTLC Calculations		Lower TTLC Calculations	
Scenario Modeled	Residents Near Landfill	Waste Worker	Residents on Converted Land	Ecological
Organic chemicals	CalTOX landfill	PEA worker organic [Work_org.xls]	CalTOX land conversion	Described in DTSC (1998a)
Inorganic lead	LeadSpread off-site [Offsite.xls]	LeadSpread worker [Worker.xls]	LeadSpread land conversion [Landconv.xls]	
Inorganic chemicals	PEA off-site [Off_risk.xls] [Off_haz.xls]	PEA worker inorganic [Work_met.xls]	PEA land conversion [Lnd_risk.xls] [Lnd_haz.xls]	

[a]Source: Adapted from DTSC (1998a).

and further data summarized in a separate spreadsheet "datcal.xls" maintained by DTSC.

The preliminary endangerment assessment model (PEA) (DTSC 1998a, p. 794) was specifically modified for application by DTSC and also encoded in Microsoft Excel spreadsheets. DTSC modified a basic spreadsheet with specifics for the particular exposure scenario, chemical class, or desired toxicity measure (hazard index or lifetime risk estimate) to yield six spreadsheets. LeadSpread (DTSC 1998a, p. 775) is a lead risk-assessment model maintained by DTSC to estimate blood lead levels in adults and children due to exposures to lead-contaminated environmental media. For this application, it was implemented in a Microsoft Excel spreadsheet. As with the PEA, the spreadsheet was modified for each exposure scenario to contain information specific to that scenario.

The modeling effort for the ecological scenario was considerably more limited. No implementations equivalent to the spreadsheets used for human receptors in the exposure scenarios were applied for ecological receptors.

ANALYSIS OF SCENARIOS AND MODELING

Because of the large number of assumptions that went into the exposure scenarios, it is not feasible to address them all in this report. The committee was not asked to conduct a complete review of the extensive modeling; it did look in detail at selected areas. In both the scenarios and modeling, there were substantial errors or omissions in conceptualization or application of the models. In this chapter, the apparent reasons for the shortcomings are examined by illustrating a few specific concerns. A more extensive discussion of the details of particular modeling problems is given in Chapter 4. The scenarios and modeling will be examined according to the requirements discussed in Chapter 2. The following discussion examines the adequacy of DTSC's proposal with respect to

- *Scenarios: Connection to Policy.* How the scenarios used by DTSC are selected, and how they are connected to the policy aims of the program.
- *Scenarios: Completeness and Coverage.* How it was ascertained that the scenarios examined are necessary and sufficient.

SCENARIO SELECTION AND MODELING

- *Scenarios: Physical Processes and Models.* What physical processes act to disperse contaminants from waste within each of the various scenarios, and how those physical processes are modeled within each scenario.
- *Mathematical Models and Their Implementation.* The mathematical representation of the physical models used within each scenario, how those models are simplified, and how the simplified mathematical models are implemented.
- *Parameter Values Used.* How representative are the parameter values selected by DTSC and used in the models for each scenario?
- *Variability and Uncertainty.* How well do the parameter values account for variability across the State of California (with respect to geographical location), and between individuals within California? How the uncertainty of some of the parameter values is taken into account.
- *Sensitivity Analyses.* The adequacy and completeness of the sensitivity analyses performed by DTSC.
- *Validation and Quality Control.* The adequacy of the procedures adopted by DTSC to ensure validation and quality control of its process, and the effectiveness of those procedures.

The ecological scenario is considered separately, because DTSC implemented it in a very different manner from the others.

The committee does not purport to provide an exhaustive list of errors, omissions, or limitations of the modeled scenarios; rather it illustrates several examples of particularly egregious difficulties that were encountered. It is incumbent upon DTSC to review all the scenario and model assumptions, replace the ones that most clearly are in need of modification, and evaluate the assumptions in light of newer data and more realistic scenarios.

Scenarios: Connection to Policy

As discussed in Chapter 2, the relevant policy objectives for this program have not been stated in adequate detail. Without any discussion of the policies that the DTSC wishes to implement, the committee found great difficulty in assessing the appropriateness of the exposure scenarios selected by DTSC. The DTSC in written and oral responses (DTSC, personal commun., October 9, 1998, see Appendix C, No. 28; DTSC, personal commun., November 20, 1998) suggested that there had been some sort of process by which the scenarios included were selected; however, the process was not documented.

An illustration of the difficulty in examining the appropriateness of the scenarios arises in the waste-worker scenario used to develop the upper TTLCs. DTSC has apparently decided to regulate some wastes based on the typical behavior observed for workers at landfills (a scenario in which institutional control or health and safety rules are absent). However, the committee assumes that DTSC has some influence on the behavior of such workers through direct regulation of the landfills themselves. Did DTSC consider, under the California Regulatory Structure Update (RSU), the possibility of modifying landfill worker behavior to meet required goals for landfills, rather than modifying the allowed input to landfills? There is no documentation that such an action was ruled out as a policy; but if not, then the waste worker scenario selected might be completely inappropriate—a better approach might be to select acceptable waste concentrations based on other criteria, and then regulate the behavior of the landfill worker, as is done for hazardous wastes.

Many of the details of the adjacent resident scenario were heavily criticized in the public comments. Those comments suggested that it is implausible that the scenario as developed by DTSC could occur, and the committee agrees with that criticism (for more details, see Chapter 4). Indeed, some extreme examples cited in the comments (e.g., the quarter-acre plot with cows and chickens, a large garden, and a fishing pond, all simultaneously used for provision of food) are literally impossible and cannot be construed to represent a plausible (let alone likely) scenario for adjacent residents. This scenario raises serious questions about both the selection of parameter values (see below) and the relationship of the scenario to policy goals. For example, a scenario such as this might be appropriate if the policy goal is the protection of the most-exposed individual, but if the scenario is meant to be plausible the parameter values must be chosen so that they are not mutually exclusive (even at the extremes of distributions of such parameter values). When asked about this particular example, DTSC stated that only one pathway almost always dominates any exposure scenario (DTSC, personal commun., November 20, 1998); however, this response is not sufficient to reduce the criticism of this scenario without more information from DTSC on how it arrived at this conclusion.

Although the scenarios discussed above for the calculation of TTLCs are flawed, they are explicitly defined. However, the derivation of soluble or extractable regulatory thresholds (SERTs) does not appear to fit into any of them; indeed, no formal definition is given of the precise scenario used for SERT derivation. The scenario appears to be based on drinking water ingestion, and may be loosely related to the adjacent resident scenario. However, any correspondence between the scenarios is not documented, and there appear to be significant differences—for example the raininess implicit in the SERT scenario appears to contradict the dustiness of the adjacent resident scenario. Once again, the major problem appears to be poorly defined policy goals which, according to oral presentations by DTSC (DTSC, personal commun., September 10, 1998), were changed during the process. However, neither the original nor subsequent policy goals are documented. The SERT scenario as implemented appears to be based more on a worst-case analysis than any other scenarios because a worst-case landfill, as well as a worst-case location for residents living downgradient of this landfill, was used. Such conditions might not actually exist in California. The selection of a single dilution attenuation factor of 100 to generically represent California landfills has not been justified, particularly in a probabilistic scheme. Moreover, DTSC has made no attempt to connect California landfill conditions to the generic landfill conditions used by the U.S. Environmental Protection Agency in its original development of the RCRA toxicity characteristic leaching procedure (TCLP) (not cited by DTSC in its documentation). The derivation of the liner protection factors is misplaced, in that it starts from a hypothetical base-case that does not (as far as the committee can tell) correspond to the U.S. Environmental Protection Agency's generic base-case. And there is no explained connection between the derived criteria and the toxicity indicators used (surface-water-quality-criteria, maximum concentration limits and the calculated groundwater concentrations). For example, is it California policy that all groundwater must meet maximum contaminant level requirements, or surface-water-quality criteria? Logically, if such a connection exists, it should exist at all downgradient distances, not just those corresponding to a dilution attenuation factor of 100 (see Chapter 4 for a further discussion of dilution attenuation factor).

It is incumbent on DTSC to devise and document meaningful scenarios. The choice of the adjacent resident scenario, the waste worker scenario, or the land conversion scenario depends to a great extent on policy aims, and their effective application depends strongly on the details of their implementation. Although the general types of scenarios selected might be appropriate, DTSC needs documentation of the process of scenario selection, and the specifics of these scenarios almost certainly will need to be modified.

Scenarios: Completeness and Coverage

DTSC has not provided sufficient information on all the types of waste situations that it is trying to address. For example, if the scenarios involving landfills are supposed to also take account of waste piles (as suggested by DTSC), this approach must be justified explicitly. The committee questions whether the approach taken by DTSC has resulted in some of the peculiar selections of parameter values, or has resulted in combinations of parameter values (each selected from possible situations) that are used to construct physically impossible scenarios. For example, 100% of the landfill surface is assumed to consist of a waste. Is that because this scenario is meant to also represent a waste pile, or because there is a policy (unstated in the scenario descriptions, but suggested in oral presentations) that some particular class of waste can be used as landfill-cover material?

Without some discussion of the totality of situations that the DTSC expects to regulate, there is no way to judge the completeness of the scenario selection or the adequacy of the few scenarios selected to take account of other situations. For example, the committee was unable to discern whether DTSC intended to evaluate potentially contaminated soil removed from one site and emplaced on another. The scenarios have to provide complete coverage for the policy goals. For example, if protection of sensitive subpopulations such as children is among the policy goals, it would be up to DTSC to demonstrate how its chosen scenarios were protective of that subpopulation. In this particular case, would the chosen scenarios adequately account for a school built on land on which contaminated soil had been emplaced?

Scenarios: Physical Processes and Models

For both health- and ecological-effects modeling, DTSC needs to provide a clear description of the scenarios to be considered, followed by an analysis of the physical processes that can occur to a waste and the individual chemicals in it under each scenario. The processes of most interest generally involve dispersion of chemicals from the waste into the environment, where receptors can be exposed to them. Such processes include evaporation of chemicals, followed by dispersion through the air; dissolution in water and transport by water flow or dispersion, adherence to dust, followed by dispersion of that dust by the wind; adherence of contaminated soil to a person's skin, followed by diffusion of the chemical through the skin; and many others.

The object of such descriptions is to elucidate the pathways that might lead to exposure of the chosen receptors. Once the pathways and processes have been identified, various models can be used to estimate the rate at which chemicals travel through the pathway, and, hence, the rates at which receptors would be exposed. The fate and transport models are used to evaluate the flows of chemicals through the environment under the conditions envisioned by the scenarios. To that end, the correct models have to be selected to represent the physical processes involved in the scenarios; and parameter values appropriate to the scenario must be used.

For example, in the landfill scenarios, little consideration has been given to the groundwater pathway (although the groundwater pathway is discussed separately in the SERT derivations, it should also be incorporated in any multipathway assessment); wind-blown dust is believed to be the predominant source for exposure pathways. There are multiple problems with the assumptions selected for evaluation of these pathways, ranging from assumptions about mixing-height to particle-size distributions to the assumption of a monofill dump. Each of these problematic assumptions affects the expected exposure, yet the specifics regarding their selection and use are not clearly identified.

An example of inadequate specification of the physical processes occurs in the modeling of dust exposures. Exposures to dust occur for all the modeling scenarios for humans: the on-site landfill worker, the off-site resident near a landfill; and the resident on converted land. In all cases, the DTSC has arbitrarily selected a dust concentration of 50 micrograms per cubic meter ($\mu g/m^3$) to represent the airborne dust at a contaminated area because it corresponds to a typical ambient average dust concentration. Such a choice is both incorrect and internally inconsistent. For example, the off-site resident near the landfill is supposed to be exposed to 50 $\mu g/m^3$ from the adjacent landfill, whereas the resident on converted land is exposed to 50 $\mu g/m^3$ from land on his or her property. Thus, two distinct, small source areas each contribute concentrations equal to this total.

An adequate representation of total exposures to dust requires three modeling efforts: an emission-rate estimate, a dispersion estimate, and a deposition-rate estimate. The first can be accomplished by using dust-emission models (e.g., those available in EPA's AP-42, 1997b), the second can be accomplished using standard dispersion models (e.g., industrial source complex (ISC) model, fugitive dust model (FDM)), and the third can be accomplished using standard models (e.g., the California Air Resources Board (CARB) algorithms, as implemented in ISC2 (EPA 1992); or the acid deposition and oxidant model (ADOM) algorithms, as implemented in ISC3 (EPA 1995b)).

Mathematical Models and Their Implementation

Estimation of environmental concentrations resulting from particular physical processes is generally carried out by using mathematical models. In principle, other approaches could be used (e.g., using analog models or direct measurement in matching situations), but the alternatives are generally too limited, too expensive, or otherwise impractical. As mentioned above, the particular models used must match the processes involved in the particular scenarios selected. Unfortunately, there are many instances in which this does not occur in the DTSC modeling. The organic vapor emission model is one example, but there are other mismatches (see the discussion of dust modeling above and the discussions in Chapter 4).

The organic-vapor-emission model used in the waste worker exposure spreadsheet is incorrect for the scenario described. The adopted model corresponds to the average emission rate over some period if a uniform contaminated material is placed (to infinite depth) over the whole area (in this case, the whole landfill) at time zero, with no infiltration rate of rainfall. For a landfill, contaminated material is unlikely to be placed uniformly to great depth over the whole landfill just at the time of initial employment of any individual worker. Although a zero infiltration rate is possible in some areas of California, organic vapor emissions are strongly affected by rainfall infiltration in wet climates, or by the effective negative infiltration in dry climates. Omission of such an important physical process as positive and negative infiltration cannot be justified.

The organic-vapor-emission model used has two principal problems even for situations of zero infiltration for rainfall. For highly volatile materials, the predicted emission rates are so high that any practical depth of contamination is rapidly depleted in its entirety. Although not so obvious, the emission-rate predictions for highly involatile materials are also much too high. Such involatile materials hardly evaporate at all, and so are barely depleted. In practice, their emission rate is governed more by diffusion through the boundary layer of air above the soil than by diffusion through the soil—the depletion depth is measured in microns. Thus two further physical processes have been omitted—depletion of the source material and the presence of an important evaporation barrier.

Moreover, even the (incorrect) organic-vapor-emission model selected was incorrectly implemented in the spreadsheet, because an additional factor of 0.1 was included. No justification for such a factor was documented, and there is no physical process to which it might correspond.

Scenarios and Models: Parameter Values

Even if adequate mathematical models are selected to match the physical processes occurring in each scenario, the evaluations can be invalidated by selection of incorrect parameter values for use in those models. Chapter 4 discusses many cases of incorrect or unjustified parameter values. Two examples of one class of error in the selection of parameter values that appears to be fairly pervasive in DTSC's modeling are given here. This error is the selection of parameter values that do not correspond to what is required for the scenario.

In the modeling of dust exposures, the deposition rate used by DTSC is said to come from a report done by D. P. Hsieh and co-workers in 1996 (provided in DTSC 1998a, p. 555). That reference does not contain any estimate for deposition velocity for wind-blown dust. It does instead have a deposition velocity, and a standard deviation (SD) for that velocity, for air particles, based on many measurements of deposition rates for ambient particles under many environmental conditions. The numerical values for the mean and SD used by DTSC are actually very different from those in the cited reference, but this transcription error is not the issue here. There are two fundamental errors involved in using the cited reference for the deposition velocity in DTSC's modeling of the dust-deposition components of the scenarios. First, deposition velocity depends on particle size and environmental conditions (e.g., wind speed, surface roughness), so that the particle-size distribution for the dust in DTSC's scenarios is important. The size distribution for background ambient particles is substantially different from that for wind-blown dust, so that the measurements do not correspond to the sizes of the particles involved. Second, DTSC used the standard deviation of all the measurements (due to geographical, temporal, surface condition, and meteorological variation), but this value does not correspond to what is required in the scenarios, which is the variability (from place to place) of the long-term average deposition velocity.

A similar problem with a variability estimate can be found in the Hsieh report (provided in DTSC 1998a, pp. 555 ff.) estimate for average ambient dust concentration. The dust modeling performed by DTSC is incorrect, because the mean and variance (from place to place in California) of the long-term average ambient dust concentration is required. However, the values given in DTSC (1998a, pp. 555ff) clearly do not correspond to such annual averages. What appears to have been given is the mean and SD of all the daily measurements at 42 stations, instead of the mean and SD of the annual averages at those 42 stations.

The air dispersion modeling used in the modified PEA and Lead-Spread models was also inadequate, principally as a result of selection of an incorrect parameter value. This modeling purported to use a box model; yet the vertical dimension of the box is not discussed. Tracking down the reference to the value of this parameter leads to the PEA documentation (DTSC 1998a, p. 794), where again this parameter value is not discussed, but is simply selected as 2 m. A box model might be adequate in both contexts; but the parameter values are critical and must be based on physical processes, not arbitrarily selected. In the original context of the PEA model, an urban garden of a dimension about 22 m, a box height of 2 m might be appropriate; but it is certainly not appropriate for DTSC's dispersion modeling of a landfill, where the dimension of the landfill is approximately 670 m. Even for such a simple model, some discussion of the physical principles involved is required.

Examination of the support documents used for CalTOX indicate that there are other problems with DTSC's parameter value estimation. Two such problems can be found in the estimate for the partition coefficient for plant tissue for trichloroethylene (TCE) (Hsieh et al. 1994, p. 25). First, this parameter is estimated for TCE based on a regression equation computed by Travis and Arms (1988) using 29 persistent organochlorines. What is not mentioned, and what is likely to completely invalidate the use of such data, is that TCE has a much higher vapor pressure than any of those 29 chemicals. In other words, the correlation is being used far outside its likely range of validity. Second, a further problem is that the estimated value is obtained from a log-log correlation, with the assumption of a normal error distribution on the logarithmic scale; but the estimated value is then reported as the mean in Hsieh et al. (1994). In such circumstances, the value estimated from the correlation is a biased median estimate, not a mean.

These two errors appear to be pervasive—the same type of errors occur for several of the other biotransfer factors listed in Hsieh et al. (1994). For biotransfer factors to meat, milk, and eggs, correlations derived for compounds that are practically not metabolized are being extended to compounds that are rapidly metabolized. Once again, the correlations are being used far outside their range of validity.

Models: Variability and Uncertainty

The treatment of variability and uncertainty[1] is inadequate in the DTSC report. No distinction is drawn between the two. Such a distinction is important in establishing compliance with health protection goals such

---

[1]The definitions of variability and uncertainty given below are taken from the NRC report *Science and Judgment in Risk Assessment* (1994).

Variability is defined as the individual-to-individual differences in quantities associated with predicted risk such as in measures of or parameters used to model ambient concentration, uptake or exposure per unit ambient concentration, biologically effective does per unit exposure, and increased risk per unit effective dose.

Uncertainty is defined as the lack of precise knowledge as to what the truth is, whether qualitative or quantitative. Uncertainty may occur in estimates or the types, probability, and magnitude of effects of and/or exposures to a chemical.

as protecting 95% of the population 90% of the time, although it is difficult for the committee to comment extensively on this topic, because of the lack of adequate documentation of the policy aims to be met by DTSC's modeling.

For practically all distributions, DTSC has assumed lognormality. (There are a few exceptions provided in DTSC's response to the committee—although one of the distributions that was supposed to be triangular was inadvertently omitted from the spreadsheet implementation.) DTSC has not verified the adequacy of this assumption, and the committee cannot confirm it. It is plausible that such an assumption might be adequate for certain calculations, for example, for estimating the standard deviation of the result of the calculations. However, to the extent that policy was articulated, it called for regulations based on the 90th percentile of output distributions. Some attempt needs to be made to verify that the distribution of results is not too affected by the assumption that all input distributions are lognormal, particularly where very few distributions contribute the majority of the variability in the output.

Models: Sensitivity Analyses

Sensitivity analysis is a procedure for determining how sensitive the results of a complex model or analysis are to various assumptions about the value of parameters and the structure of relationships. The construction of complex analyses, such as those used by DTSC in the proposed waste-classification system, should be accompanied by an extensive and comprehensive strategy for analyzing the sensitivity of the results of the analyses. Although the model is being developed, the repeated application of sensitivity-analysis techniques allows the developers, the reviewers, and, ultimately, the users of the model to do the following:

- Identify the dominant factors that most influence the models' results.
- Focus attention on assumptions concerning dominant pathways and improve efforts to reduce the uncertainty associated with them.
- Focus development, analytical, and review efforts on the most important assumptions on which the model and analysis are based.

A comprehensive sensitivity analysis requires a full understanding of the structure of a model and the nature and strength of the multifarious assumptions that have been made in its construction. It is rarely possible to conduct a sensitivity analysis on every assumption a model contains, and distilling useful insights from such a process would be almost as difficult as undertaking the analysis itself. Thus the sensitivity analysis must, in its design, include substantial insight into the factors that have the potential to most influence the results.

To the extent that sensitivity analyses were conducted, no apparent effort was made to communicate the insights gained from these analyses to the public in a clear manner. The statistical results were summarized deep within the appendices of the reports presented for review. It requires a reader with substantial experience in stochastic models to interpret the results that were summarized. It might be that this failure to communicate the results in a clear manner resulted in substantial unnecessary debate about assumptions that were not important. For example, DTSC staff responsible for some of the analyses indicated that the sensitivity analyses demonstrated that the model's results were insensitive to assumptions about how much clothing a landfill operator was wearing because dermal exposures were not responsible for significant amounts of risk. Assuming that this interpretation is correct, communicating it clearly to the public would have saved significant amounts of controversy.[2]

Another method of assessing the sensitivity and quality of the models is to evaluate intermediate results within the calculation. DTSC should develop basic model input data and make available all intermediate output data for independent evaluation. This would allow an external group, such as the NRC committee, to evaluate the results of the modeling more completely.

At the second public meeting, DTSC staff members suggested that the risk estimate obtained from the scenario selected is typically dominated by exposure via a single environmental pathway. They further asserted that the scenarios selected account for all of the important environmental pathways and should suffice to protect the health and safety of California residents. The documents presented to the committee do not provide adequate support for either of these contentions.

---

[2]The fact that one of the questions DTSC submitted to the committee focused on the dress habits of landfill operators suggests that the results of the analysis were not clearly communicated to the department managers either.

Specifically, detailed sensitivity analyses and comparative modeling runs of selected sites should be included in the document.

DTSC used the sensitivity analysis options currently available in the Monte Carlo software, Crystal Ball by Decisioneering, Inc. Although such subfunctions are quite useful for a scoping analysis, a more thorough treatment of sensitivity analysis is required. In particular, the only parameter values for which sensitivity analyses were performed were those assigned distributions in the analyses. However, some of the most important parameter values (e.g., soil-mixing depth, dispersion model box height) were not assigned distributions, but were simply given as point estimates. These are usually the assumptions that are most uncertain and based on the least knowledge. The sensitivity of the results to such parameter values was not examined in detail, if at all.

Moreover, some of the parameter values that ought to be included in a sensitivity analysis are hidden in the support documents. For example, the half-lives of chemicals in the ground-surface and root-zone soils (Hsieh et al., 1994, p. 22) are based on estimated values (reported as measurements), but it is arbitrarily assumed that there is an extra factor of 5 uncertainty. The sensitivity of any of DTSC's results to this factor is unknown because sensitivity analyses were not performed on any of the uncertainty factors. Upper percentiles of the risk distribution might be very sensitive to such factors.

A detailed analysis of variance of all results (see, e.g., McKone and Ryan 1989), coupled with other stepwise regression analyses might lead to a different conclusion as to which parameters are most influential. For example, variables possessing considerable colinearity could be reduced in their apparent significance. The committee cannot fully evaluate the strengths and weaknesses of DTSC's approach without such analyses.

Even within the framework of the sensitivity analyses performed by DTSC, it would be useful to see the results for several selected cases. This could permit assessment of questions such as: Is it always a single pathway that dominates? How do the exposures from the various scenarios differ? If multiple pathways contribute, how are they best controlled? DTSC documents do not appear to provide this information.

Models: Validation and Quality Control

There appears to have been very little validation of any of the models

SCENARIO SELECTION AND MODELING 73

used by DTSC, in any sense of the term. In the context of DTSC's approach, validation could include the following:

- Each exposure scenario selected should correspond to a real-world situation.
- Each physical process that is modeled should occur in the corresponding scenario.
- The theoretical models describing the physical processes should correctly represent such physical processes.
- The simplified mathematical models representing the theoretical models should adequately approximate the theoretical model over all allowed ranges of input values.
- The implementation of the model should correspond to the simplified mathematical model; that is, the implementation should be a mapping of the simplified mathematical model (expressed in mathematical notation) to a computational scheme (e.g., a spreadsheet, a specialized computer program, or a set of such programs) in such a way as to preserve the mathematical structure of the simplified mathematical model. The only difference between numerical results produced by the implementation and those produced by the simplified mathematical model, for any allowed input values, should be due to the finite arithmetic precision of the practical computational devices. Good practice should be applied to minimize any such differences.
- The parameter values used in running the implementation for any particular scenario must be correct (i.e., match experiments) for the conditions of that scenario and should correspond to the definition of the parameter in the mathematical model (i.e., be based on measurements of the correct parameter).
- The results from the implementation of the model should be acceptably close to observations over the entire range of input values.

Although it is unlikely that all these validations would ever be fully performed for any modeling system, it is quite apparent that many of them that are entirely feasible have not been carried out for many of the model implementations used by DTSC. Thus, the committee has found the following disparities:

- The physical processes occurring in the scenarios differ from the models selected to represent them (dust emissions, dispersion of dusts and vapors, vapor emissions).

- The implementations of some models do not correspond to the simplified mathematical models described in the documentation.
- Some input values do not correspond to the values required by the model in the context of the particular scenario.
- Some input values used differ from the documentation.

ECOLOGICAL SCENARIO

The approach taken to ecological evaluation is substantially different from that adopted for evaluation of human health risks. DTSC proposed a two-step method to determine the potential level of hazard of wastes to wildlife. In DTSC's report, this approach was adopted "due to the lack of an accepted risk assessment methodology for ecological toxicity (DTSC 1998a, p. 62)." This methodology was followed only for the lower TTLCs, that distinguish nonhazardous from special wastes, because it was thought that the protection of ecological resources would be similar in class I, class II, and class III landfills. No justification for making this assumption was given by DTSC. DTSC should provide a rigorous validation of this assumption.

The first step was the use of the values presented in the federal proposed HWIR (60 Fed. Register 66344, Dec. 21, 1995). The HWIR methods include scenarios developed for residential exposure and a variety of ecological effects, although the similarity of any of these scenarios to those adopted by DTSC is not discussed. The documents describing the HWIR approach are obscure (RTI 1995a,b), and were not provided to the committee for review. It is unclear to what extent wildlife-specific exposure scenarios were adopted in the HWIR.

DTSC did develop ecological-effects TTLCs for those chemicals for which the HWIR analysis indicated that the derived exit-level (similar to a TTLC) for ecological receptors would be less than the exit level for humans using the HWIR scenarios and analyses. The methods applied by DTSC to derive TTLCs that are specific for wildlife did not include wildlife-specific exposure scenarios, but rather only applied wildlife-specific threshold values for the hazard of chemicals. Thus, the lower threshold TTLCs derived were not truly specific to wildlife, because they still relied on a scenario of human exposure. This approach might not always be protective of wildlife for several reasons. The diets of wildlife are often not as varied as those of humans. Wildlife eat what is available in their home range and, thus, can have significantly greater exposures than humans. A good example of this is birds living in areas adjacent to hazardous-waste landfills along the shores of the Great Lakes (Ludwig et al. 1993). In those areas the exposure of wildlife is much greater than that of humans, and because the toxicity of chemicals might be similar in wildlife and humans, the risk to wildlife could be much greater than that to humans. Based on this line of reasoning, the use of human exposure scenarios is inappropriate for wildlife.

Of the 36 chemicals listed by DTSC, 28 have both HWIR human or HWIR ecological exit levels[3] (DTSC 1998a, p. 63). Of the 28 chemicals, 11 have HWIR ecological exit levels were less than HWIR human exit levels[4], and 17 have higher values (DTSC claims 18 of 29, see footnote 3). Based on this analysis, DTSC argued that the application of lower threshold TTLCs derived from human health endpoints and human exposure scenarios would be protective of ecological receptors for those 17 (or 18) chemicals. Of the eight chemicals for which no HWIR comparison was possible, DTSC contended that "the other eight chemicals [for which no HWIR ecological exits levels were calculated] were considered low priority in the HWIR analysis and, therefore, are not likely to present significant threats to ecosystems at concentrations below those that would be of concern for human health."

Although either or both of these arguments may be true, DTSC provides no analytical support for these conclusions. In fact, because HWIR exit levels for ecological receptors were less than human receptors for approximately one-third of the chemicals, it could be argued that the application of exit-level TTLCs based on protection of human health would fail approximately one-third of the time (but see footnote 4). This seems to be a strong argument for the development of a specific ecological risk-assessment scheme for wildlife receptors.

For hexavalent chromium, one of the 11 chemicals passing thought the HWIR screen, DTSC indicated that the proposed human-health-based TTLC equaled the HWIR ecological exit level, and considered this to demonstrate that the TTLC would be adequately protective of wildlife.

---

[3]In the text, DTSC (1998a, p. 62) claims that 29 of the 37 chemicals being analyzed had both HWIR human and HWIR ecological exit levels; and at other parts of the report a reference is made to 38 chemicals being analyzed.

[4]The committee notes that the HWIR implementation contained many errors, so that it is impossible to confirm that the same results would be obtained from a correct implementation of the intended HWIR approach.

Again, analytical support is lacking. For the remaining 10 chemicals for which the HWIR ecological exit levels were lower than the human exit levels (endrin, methoxychlor, lead, mercury, selenium, nickel, vanadium, cadmium, zinc, and copper), DTSC moved to a second step of ecological risk assessment. The development of lower-threshold TTLCs for these chemicals was based on an additional chemical-by-chemical analysis by DTSC. This further analysis was primarily a refinement of the reference doses for wildlife and did not include a refinement of the exposure assessment. The lower-threshold TTLCs derived included consideration of background concentrations and screening-level values derived by other organizations. The TTLC values derived for the 10 chemicals seem reasonable, but were not derived by use of a defined wildlife-specific ecological risk-assessment procedure. Because this refinement only considered the hazard portion of the risk assessment, it is inadequate to demonstrate protection of wildlife. Furthermore, the methods applied to the 10 chemicals are somewhat arbitrary and chemical specific, and could not be applied as a more generalized method for other chemicals.

DTSC has no explicit plan or method to add more chemicals to the list of TTLCs. Currently, the ecological methodology is at best incomplete and poorly justified, and at worst potentially underprotective. With respect to the ecological methodology used by HWIR, the committee agrees with the U.S. Environmental Protection Agency's Science Advisory Board (EPA 1996):

> The ecological analysis in the HWIR document is fundamentally flawed because a lack of data has been implicitly equated with lack of adverse ecological effect throughout the analysis. As a result, only a handful of well-studied chemicals have actually received a scientifically credible review. The Subcommittee recommends, therefore, that the Agency discard the proposed screening procedure for selecting the initial subset of chemicals for ecological analysis and instead require that a minimum dataset be satisfied before ecologically based exit criteria are calculated. For those chemicals for which the minimum dataset cannot be satisfied, the Agency should clearly indicate that the exit criteria are based solely on human health considerations. The exit criteria should be re-evaluated, however, when and if additional data on ecological effects become available.

4

Issues of Model Application

AFTER CONSIDERATION of the scenarios and models chosen by the Department of Toxic Substances Control (DTSC) of the California Environmental Protection Agency in Chapters 2 and 3, the NRC committee examined the quality of the data used in the models for classification of wastes. This chapter examines (1) detailed model parameters for exposure pathways such as those leading to dietary intake; (2) selection of parameters, such as dilution attenuation factors, for specific models; (3) analytical methods, especially the use of either the waste extraction test (WET) or the toxic characteristic leaching procedure (TCLP) extraction methods; and (4) human and ecological toxicity tests.

MODEL PARAMETERS

After the selection of scenarios and models to be used in the risk assessment, the models must then be implemented with the correct parameter values. It is incumbent on DTSC to review its modeling to ensure correct selection of parameter values to correspond to the scenario in which the model is used. The committee examined certain parts of the documentation and spreadsheets to evaluate whether a suitable quality-control process had been applied to DTSC's modeling (see Chapter 3). In its review of the DTSC report, the committee found numerous errors and inconsistencies in the selection of the component models and the model parameters. The following list indicates some of the types of errors and problems that were found for input parameters. Given the nature of the task and the time allotted, the committee identified as many specific problems as it could and provides examples of these in this chapter. However, the committee did not prioritize these problems and notes that not all of them are equally serious or have the same impact on the outcome of the risk assessments. This list should be taken not as a complete set of problems that needs to be corrected, but as an illustration of the type of problems and errors that a complete quality-control program should be designed to locate and correct.

The problems and errors can be classified into several types, with some problems and errors occurring simultaneously:

- *Transcription errors*: The values have been incorrectly transcribed from an original reference.
- *Mistaken identity*: The values are correctly derived from measurements, but the measurements are of the wrong physical quantity in the context of the particular scenario and model.
- *Mistaken derivation*: The values are derived from measurements of the correct physical parameter, but the derivation is incorrect in the context of the particular scenario and model.
- *Incorrect extrapolation*: The values are derived from physical measurements by using an extrapolation that is inapplicable.
- *Impossible*: The values used are physically impossible.

For completeness, the following types of problems that should be corrected by an adequate quality-control process are also discussed in this chapter:

- *Inappropriate model errors*: The model used does not correspond to the physical processes occurring in the scenario
- *Structural errors*: Errors or ambiguities in the structure of the models that lead to errors in calculations.
- *Documentation errors*: The description of the model differs from the model that was intended to be adopted.
- *Implementation errors*: The implementation of the model differs from the mathematical model adopted.
- *Calculation errors*: Something has been incorrectly calculated, but it is not possible to determine what went wrong.

ISSUES OF MODEL APPLICATION

Of course, documentation errors and implementation errors often occur together, and either or both can occur at any stage in the translation from physical description to simplified physical description to mathematical model to simplified mathematical model to implementation of the model.

Parameter Selection for Scenarios

The most basic level in scenario development is the selection of the specific parameters needed to implement the models in the context of the scenario. Such parameters include food intakes, quantities of soil eaten, dust-deposition rates, bioconcentration factors, soil-mixing depths, vapor pressures, soil porosities, inhalation rates, and solubilities. Below is an analysis of the types of problems encountered when DTSC's choices of parameter values were examined.

Food Intake

The food intake values used in the scenarios are based on data that may not be directly relevant to the citizens of California. It may reasonably be expected that the scenarios outlined would result in a small number of individuals incurring a large, albeit unknown, risk. However, the number of such individuals relative to the population of California, and the risk incurred by these individuals, is not knowable without completing model runs using the exposure scenarios developed by DTSC. The committee is, therefore, unable to estimate the effect the food intake and population assumptions have on total risk. Some of the committee's concerns with the application of food intake data are described below.

The specific dietary intake parameter values need to be realistic. Those used by DTSC do not appear to have been selected for real-life conditions and draw upon data that are 10 to 30 years old (an example of the mistaken identity error). Focusing on the adjacent resident scenario, current data need to be gathered on the types of residents near facilities. What fraction are farm households? If they are not farm households, do these households produce and consume their own meat, eggs, and dairy products?

If DTSC selects farms as the basis for its adjacent resident scenario, it should collect data on the number of small, family-owned farms in

California because residents of these farms are most likely to use homegrown crops as principal sources of food. Are they located near waste sites? How many? How far? Are there California demographic data to support those given in the U.S. Environmental Protection Agency's (EPA) *Exposure Factors Handbook* (EFH) (EPA 1990a)? As with the discussion of the scenarios for the population subject to the food intake assumptions mentioned above, these questions also raise the problem of estimating the changes in population living near hazardous waste sites and producing their own food. With the changes in farming from small, family-owned farms to agribusiness, the risk to a small number of individuals may be reduced in time. However, the large, agribusiness farm subjected to contamination of crops by a nearby hazardous waste site could increase the risk (by a smaller amount) to a larger segment of the population through the sale of contaminated crops. Therefore, DTSC might also, in the definition of its scenarios, wish to take account of time trends in agriculture, perhaps resulting in fewer small farms near waste sites, and perhaps resulting in wider dispersion of contaminated produce from larger farms.

Some of the difficulty in the exposure assessments for the adjacent resident scenario can be traced to estimates of dietary intake, most specifically for home-grown foods. These estimates are presented in the CalTOX parameter values section of the DTSC report (1998a; pp. 611 ff). The primary reference for home-grown food intake is from the first revised EFH (EPA 1990a). Table 4-1 reports the fraction of various foods that are assumed home-grown.

TABLE 4-1 Consumption of Home-Grown Foods

Home-Grown Food Type	Fraction of Food Obtained from Home-Grown Source	
	Mean	Coefficient of Variation
Fruits and Vegetables	0.24	0.7
Grains	0.12	0.7
Milk	0.4	0.7
Meat	0.44	0.5
Eggs	0.4	0.7
Fish	0.7	0.3

Source: Adapted from Table III, Activity patterns, household parameters, and other exposure factors (DTSC 1998a, p. 613), which was adapted from EFH (EPA 1990a).

ISSUES OF MODEL APPLICATION   81

Taking even the mean values for the fraction of foods consumed would require residential conditions that would be illegal under various ordinances in most communities, and are unlikely to be observed for any individual. The problem with using these food intake estimates stems from using data collected in a specific survey and attempting to apply it to a more general or inappropriate situation. The following sections examine the problems with using the data on home-grown food intake by specific food type.

Fruits and Vegetables

The primary EFH reference for fruits and vegetables reports a decrease in the average size of the garden from 600 ft^2 in 1982 to 325 ft^2 in 1986. Extrapolation to 1999 would suggest that even smaller garden sizes are the current norm. Furthermore, the gardens are found to produce approximately 0.9 lbs of produce per square foot annually, or about 300 lbs of produce. Given the U.S. Department of Agriculture's results cited in the EFH indicating consumption of 201 g/day of vegetables, a 325 ft^2 garden would fully support two individuals on this intake of vegetables. However, this is an *average* yield and does not take into account the differential yield for different vegetables, for example, pumpkins versus spinach. The EFH further reports that the largest numbers of such gardens are in the Midwest and South and that more individuals in rural settings tend such gardens compared with those living in cities and suburban areas. Neither EFH nor the DTSC report provides specific data on the number of gardens in California. It is reasonable to assume that some state-specific data for consumption of home-grown fruits and vegetables are available for California, a major agricultural state, yet only data from the EFH are used.

Table 2-10 in the EFH shows that the percentage of home-grown fruits and vegetables consumed ranges from 4.2% for lettuce to 75% for lima beans; these values are used in the DTSC report. The EFH data cited were gathered for a specific survey and no evidence is given by DTSC to support the applicability of such data to conditions in California. The EFH specifically cautions the reader on the representativeness of these data, which were drawn from a small number of days and quantified by recall only. DTSC uses an average value taken from Table 2-10 and presumes that consumption of all fruits and vegetables matches this average value from home gardens; however, no rationale is provided for this presumption.

Furthermore, these consumption values for home-grown fruits and vegetables seem excessively large for the California population as a whole. DTSC does not specify who the target group is or who will be protected. It would appear that DTSC is looking at a maximally exposed individual. Although these values might be accurate for home gardeners in 1986, their validity for a population that consists predominantly of urban and suburban dwellers (another mistaken identity error) is questionable. DTSC has not demonstrated that the population assumed to grow and consume these foods exists. DTSC provides no support for the size of the garden versus food consumption, nor do they provide information about subpopulations who might be vegetarians, low income and subsistence farmers, specific ethnic groups, or children. Whether explicit account needs to be taken of any such subpopulations depends on the scenarios under evaluation to meet specific policy goals. The public comments indicate that it is possible to ascertain the number of hazardous-waste sites in the state and the distances of the nearest residences. With such information, it should be possible to adequately characterize home gardeners living near hazardous-waste sites, including the average distance from their residences to those sites.

Grains

For the grain consumption pathway, DTSC makes use of data exclusively on corn, because corn is the only "grain" product mentioned in the EFH. However, the committee suspects that corn grown in home gardens is used as a vegetable, not a grain (a mistaken identity error). It is not aware of any data on grinding corn meal from corn grown in a home garden. DTSC further compounds this poor data analysis in that other grains, presumably wheat and similar products, are assumed to be identical to corn. Lacking any data supporting the use of wheat as a vegetable, its use can safely be assumed to be as a grain to make flour and other products. It is extremely unlikely that the typical home garden produces 12% of the wheat flour used in the residence.

Meat, Dairy, and Eggs

The DTSC report states that, in farm households, the annual fraction of home-grown beef consumed is 44% with a coefficient of variation (CV) of 0.5 (DTSC 1998a, p. 612). Similarly, for dairy products and eggs (by direct assumption of the equivalence between dairy products and eggs), the values are 40%, with a CV of 0.7. These values might be biased because they were based, according to EFH (EPA 1990a), on a survey of 900 rural farm households published in 1966, and *they only apply to farm households*. It is highly unlikely that such numbers can apply to suburban and urban settings, where keeping livestock is usually against local ordinances (a mistaken identity error). Again, the central issue is what population is being protected? Clearly, this component of the scenario uses a value based on a maximally exposed individual, not on the broader population. With the changes in U.S. agricultural practices since the 1960's from family farms to agribusiness, the application of these data to *any* residents of California must be justified by DTSC.

Fish

The consumption rate of fish for recreational or subsistence anglers and the fraction of fish eaten from local sources are also subject to controversy. The consumption rates in the EFH are based on data collected in 1973-1974 (EPA 1990a), and might no longer be valid, particularly given the number of no-fishing advisories in effect for many California waters. DTSC also assumes that the shape of the distribution of intake is triangular, however, the shape of the triangle is not indicated and the basis for the assumption is unsubstantiated.

The committee urges DTSC to incorporate more recent exposure factors (e.g., those given in EFH published by EPA in 1997) as well as data that are representative of California urban, suburban, and rural populations.

PARAMETER SELECTION WITHIN SPECIFIC MODELS

This section highlights some of the committee's concerns regarding the use of models for soluble or extractable regulatory thresholds (SERTs) and some specific parameter problems for the preliminary endangerment assessment (PEA), LeadSpread, and CalTOX exposure models used to develop the toxicity threshold limit concentrations (TTLCs).

Soluble or Extractable Regulatory Thresholds

There are various structural problems with the DTSC's implementation process for SERTs in general:

- The use of a single dilution attenuation factor ensures that variability in the population's exposure to groundwater is not taken into account. As DTSC acknowledged during the second public meeting (DTSC, personal commun., November 20, 1998), the SERT scenario was modified for some sort of worst-case exposure, not for exposure at a 90th percentile of the population as documented.
- There is a logical disconnection between some toxicity indicators (surface-water-quality criteria, maximum contaminant level) and the calculated values with which they are compared (groundwater concentration). Logically, if such a comparison is meaningful, it should also be meaningful at all downgradient distances, not just those corresponding to a dilution attenuation factor of 100. This is probably connected to the previous problem, and its solution requires explicit specification and acknowledgment of the policy objective.

There are also problems with the SERT definition.

- The toxicity values used for the SERT calculations are particularly puzzling. These toxicity values include the ambient water quality criteria for aquatic life and maximum contaminant levels. Ambient water quality criteria apply to surface-water bodies, so that an extra dilution has to be taken into account, that is, where the groundwater runs into the surface water body. In particular cases, such as water bodies that are fed only by contaminated groundwater, the dilution factor might be greater than unity (e.g., if the water body evaporates and the contaminant is nonvolatile). If DTSC is attempting a probabilistic approach, then some distribution for this further dilution is required. If DTSC is attempting a worst-case analysis, then the worst-case would have to be applied. This type of error could be either a mistaken identity error if ambient water quality criteria were assumed to be relevant to groundwater, or an extrapolation error if it was assumed that groundwater concentrations correspond to surface-water concentrations.
- The maximum contaminant level is used as an indicator level applicable to groundwater. Given that health-based levels are separately

ISSUES OF MODEL APPLICATION 85

derived, it appears that DTSC is using the maximum contaminant level as an enforceable standard for all California groundwater. Thus, the committee questions the use of an maximum contaminant level for a risk-based approach. This appears to be a mistaken identity error.

• DTSC's approach to the use of a liner protection factor to take better account of modern landfills is also misguided. DTSC has attempted to estimate the liner protection factor by comparing a lined landfill with an unlined landfill. However, the parameter values used for the unlined landfill appear to correspond to a fairly tight landfill with clay liner. But these parameter values do not correspond to the parameter values used in the original EPA modeling. Thus, the liner protection factor is calculated from an incorrect base value. This could be classified as a mistaken identity error for all the parameter values for the unlined landfill.

• The SERT scenario is so poorly defined that the committee cannot comment on its applicability—it can simply point out where the implementation does not agree with the documentation. The following paragraphs identify some of the specific problems that were encountered in the review of the lower (nonhazardous) and upper (hazardous) SERT calculations.

Calculations for Lower SERTs

The DTSC spreadsheet for SERTs has a 100% correlation between the distributional calculations for risk and the hazard index. Although this does not affect the results of the current calculations, it is possible that in a more complex analysis such a correlation would be incorrect. (In fact, in the subsequent calculation of upper SERTs using the DTSC spreadsheet, this correlation is essential to get correct results, because the minimum function is applied at an intermediate stage of calculation). This is a potential structural error, although it does not affect the current calculations.

The logic in the spreadsheet for SERT calculations does not correspond to the description given in the DTSC documentation (DTSC 1998a, pp. 43-45). The minimum value of the health-based level, maximum contaminant level, or ambient water quality criteria is applied before the statistical lower 10th percentile is calculated for health-based level and the lower 10th percentile of this minimum has been found. In principle, this should make no difference to the final results, although one can expect problems in labeling some intermediate results (see below). This is a documentation error, or possibly a structural error (although it does not affect the final result).

Possibly as a result of the preceding calculation, the values given in the column labeled "Health-based level × 100" (DTSC 1998a, p. 46) do not correspond to the health-based level × 100, where the health-based level is computed as the 10th percentile as described in the text. In fact, for each of six chemicals (aldrin, kepone, arsenic, beryllium, thallium, and vanadium), the value given in the table is correct to one significant figure, that is, it is indeed the health-based level × 100, where the health-based level is the lower 10th percentile value. For all but four of the remaining chemicals in the table, the value given can be obtained (to one significant figure) from the same calculation, but by using the mean values of each parameter in the calculation, not the lower 10th percentile of the distribution resulting from using the parameter value distributions. So the value given does not correspond to the text description. For the remaining four chemicals (chlordane, methoxychlor, chromium VI, and molybdenum), it is not clear how the values given in the table were derived, because they do not correspond to either calculation or to anything in the spreadsheet.

The spreadsheet apparently used a Monte Carlo approach to evaluate the 10th percentile of the lognormal distribution required for calculating the health-based level for the lower SERT. Although the spreadsheet entries show correct values (within 0.7%) for the 10th percentile in most cases, in five cases (cobalt, fluoride, molybdenum, thallium, vanadium) the entries are more than 15% in error. This appears to be a calculation error. The calculation for the health-based level involves a single lognormal distribution (for the hazard index) or a multiplicative combination of three lognormals (for risk), which is also lognormal. Therefore, the calculation of the lower SERT is analytically straightforward.

Calculations for Upper SERTs

The difference between the calculation of the lower and upper SERTs is the liner protection factor. The upper SERTs are calculated by multiplying the lowest of the health-based level, the maximum contaminant level

ISSUES OF MODEL APPLICATION

or the ambient water quality criteria by a dilution attenuation factor of 100 and a liner protection factor. The DTSC documentation (DTSC 1998a, p. 47) specifies that an liner protection factor was entered as a "custom distribution" but provides no indication of how the values were derived. The custom distribution function in the Crystal Ball software by Decisioneering, Inc., allows various options for defining distributions (combinations of point values with assigned relative weights together with piecewise linear densities), but the DTSC documentation does not specify what options were used. During the first public meeting, DTSC stated that six values (two sites, three conditions, using the HELP model) were used as a custom distribution (DTSC, personal commun., September 10, 1998).

The SERT spreadsheet contains a list of six values for liner protection factors. They are (in the order listed): 36, 190, 1600, 22, 118, 970. The Crystal Ball custom distribution add-in, however, lists a different set of six values, entered as point values with equal relative weights. These values are (in increasing sequence): 18, 99, 118, 191, 970, 986. The DTSC documentation gives three generic values for liner protection factor—18, 99, 986, using an approximate model that takes into account leakage through the liner versus leakage through clay only (DTSC 1998a, p. 1,487).

The documentation later cites the HELP model as giving three values each for two precipitation and evapotranspiration regimes—Los Angeles and Eureka (DTSC 1998a, p. 1,488). The values are: 36, 190, 1,600 for Los Angeles and 22, 118, 970 for Eureka. These six values are identical to those listed in the spreadsheet, but not in the custom distribution in Crystal Ball. The documentation then appears to include a printout of a spreadsheet (source not provided by DTSC) that provides yet another set of values for all three cases: 36, 190, 1,600 for Los Angeles; 22, 120, 970 for Eureka; and 18, 99, 990 for a generic case (DTSC 1998a, p. 1,492). Thus, the documentation is not clear on what values are used to derive the upper SERTs—the resulting difference between documentation and implementation could be a documentation error or a transcription error.

The spreadsheet calculations for upper SERT results appear to correspond to the values for liner protection factors present in the Crystal Ball custom distribution (which do not correspond to any documented set of values). DTSC indicated that the spreadsheets contained the most valid calculations, so that the liner-protection-factor values in the Crystal

Ball custom distribution must presumably be taken as those to be used. Although adding in the liner-protection-factor custom distribution makes the distributions to be evaluated a sum of six lognormals, it is still relatively easy to compute analytic results—all that is required is the solution of one nonlinear equation each for the risk case and the hazard index case.

Examination of the table containing the liner protection factors (DTSC 1998a, p. 48) shows just four values: 18, 19, 24, and 63, although it is not clear how these values were derived. They are not calculated within the SERT spreadsheet. It is possible that the erroneous 19 and 24 arose from the same error that produced the >15% error in the health-based levels for the lower SERTs because the same five materials are affected (cobalt, fluoride, molybdenum, thallium, vanadium). The exact values using the liner-protection-factor set present in the Crystal Ball custom distribution should be: 18, 30.1, and 63.7. These apply when the upper SERT is based on maximum contaminant level–ambient water quality criteria, oral reference dose, and cancer potency, respectively. The value obtained when the upper SERT is based on maximum contaminant level–ambient water quality criteria is necessarily the lowest of the six-point estimates for the liner protection factor, because the liner-protection-factor distribution is the only distribution involved in this case. Each point in the liner-protection-factor distribution corresponds to one-sixth weight, so that all percentiles from 0 to 16.66666 are assigned this lowest value. It is not clear if this was DTSC's intent. For the six liner-protection-factor values specified for Los Angeles and Eureka, the three corresponding SERT values would be 22, 34.4, and 54.5. The results given by DTSC have to be considered a calculation error.

Preliminary Endangerment Assessment

A major concern for the preliminary endangerment assessment (PEA) model is that DTSC fails to provide a thorough description of the scenarios to be modeled. A review of the component models of the PEA model indicates several areas where the component models are incorrect, as indicated below. In addition, many of the model values are not referenced, or differ from the documentation cited. Furthermore, the uncertainty distributions are not always documented. Some incorrect pathway models include evaporation of chemicals from a landfill and vapor

Issues of Model Application

dispersion above a landfill, for which DTSC fails to use the available (roughly correct) pathway model in CalTOX.

Risk and Hazard Spreadsheets

For the adjacent resident and the resident on converted land scenarios, DTSC used two spreadsheets, one for hazard and one for risk. This might present a fundamental difficulty related to the definition of the level of protection desired, if the requirement is to protect each individual against particular levels of both noncancer and cancer risks at some percentile of the distribution across individuals. With the two spreadsheets, the requirements are imposed completely separately. Thus, even though 90% of individuals are protected at the selected levels against cancer risk, and 90% against noncancer risk, less than 90% (in principle, as few as 80%) could theoretically be protected against both cancer risk and noncancer risk simultaneously. This is a structural error, but DTSC's goals have not been sufficiently well defined to determine whether the structural error is potential or actual.

Exposure Pathway Factor for Inhalation

The definition of "Pathway Exposure Factor for Inhalation" (DTSC 1998a, p. 790) differs (in definition and dimensionality) from the definition in the spreadsheet. The spreadsheet definition is preferable because it maintains equal dimensionality for this quantity for all exposure routes. Based on the values used in the spreadsheet, the receptor is a child for the hazard calculation, although the precise receptor does not appear to be documented. These variations appear to be documentation errors. Only 16.3 hours of the day is accounted for in "moderate activity" and "resting" (documentation) or "resting (sleep)" and "heavy activity" (spreadsheet); it is unclear what happens to the rest of the day for this receptor. There is no truncation on the distributions used, so that the length of a day could theoretically exceed 24 hours during the Monte Carlo procedure (allowing an impossible parameter set).

Equation 8 for the pathway exposure factor for inhalation in the residents in the land conversion scenario (DTSC 1998a, p. 793) omits the effect of leaching and vapor emissions, and so is incorrect for soluble and volatile materials. In view of the length of time involved (20 years), the error introduced is substantial. This has to be considered a structural error in the model.

Dust

As mentioned in Chapter 3, DTSC's approach to dust is inadequate for the adjacent residents scenario (DTSC 1998a, p. 792). That discussion is expanded here by looking in more detail at some specific examples that illustrate this inadequacy. There are suitable models in AP-42 for dust-emission rates and for dust-deposition rates (California Air Resources Board model subroutine, as used in industrial source complex (ISC2) model (EPA 1992), or the latest models used in ISC3 (EPA 1995b). Both account for dust-particle size distributions. The fugitive dust model also might be appropriate (EPA 1990b). Using such models would readily allow evaluation of the variation of dust deposition with distance and meteorological conditions, something that is lacking from the DTSC approach. (These models are easy to construct if they are not available prepackaged.) Failure to use such modeling approaches is a structural error for this, the major pathway.

DTSC stated in its discussion of adjacent residents that "the residence is assumed to be very close to the landfill, so no dispersion of dust is included in the model" (DTSC 1998a, p. 785, Section 1.2.1). The committee does not understand this assumption because DTSC does not provide any evidence that even the short distances involved do not result in significant dispersion. After all, the residence is not in the middle of the landfill—it must be to one side of it. Thus, it can be expected that, on average, the wind is blowing in the wrong direction perhaps half the time. That amounts to a neglected factor of 2 (averaging over landfills and residences statewide), even neglecting dispersion. Furthermore, one property of dust is that it settles as a function of distance from the landfill, resulting in a steady change in the importance of the ingestion pathway as distance from the landfill increases.

At the second public meeting, DTSC indicated that a concentration of 50 $\mu g/m^3$ of dust in air (based on the National Ambient Air Quality Standards) was used as a default value and that dust emission and dispersion had not been modeled because there were no models for dust emission (DTSC, personal commun., November 20, 1998). This state-

ISSUES OF MODEL APPLICATION

ment was very disturbing to the committee, given that dust-emission models are easily accessible in AP-42 (EPA 1997b) and that any consideration of this type of scenario (adjacent resident) clearly calls for adequate modeling of the dust-dispersion pathway (the primary off-site exposure pathway considered). The selection of the National Ambient Air Quality Standard value was apparently also based (according to the information provided in the second public meeting, but not elsewhere documented) on measurements of ambient atmospheric particle loading. However, except in the vicinity of dominating local sources, the great majority of the ambient particle loading in the atmosphere is due to long-range transport. The DTSC treatment of dust can be considered an example of incorrect extrapolation—the measurements of ambient dust loads have been incorrectly extrapolated to the DTSC scenario.

Compounding the incorrect extrapolation are some transcription errors. In the DTSC report (DTSC 1998a, pp. 785, 791, 815), it is stated that the deposited dust mixes to a depth of 0.15 m over a period of 20 years. However, in the spreadsheets for the adjacent resident scenario, 0.015 m and 0.1 m are used for hazard (Off_haz.xls) and risk (Off_risk.xls), respectively. Thus, risk and hazard calculations correspond to different scenarios, and both are different from the scenario that is presented in the DTSC documentation. One of these transcription errors is propagated to the lead spreadsheet, which uses as one of its inputs a dilution factor (0.00098) evaluated in Off_risk.xls.

Dust_conc is defined as the concentration of dust in air over the landfill (DTSC 1998a, p. 792). However, what is required here is the concentration of dust over the resident, not over the landfill. The model used by DTSC should, but does not, differentiate between the landfill and off-site, an extrapolation error. The dilution due to dust dispersion and deposition should be taken into account. In addition, a more subtle point should be taken into account: only one average dust concentration is required for an on-site worker present during the day; for the off-site exposures, two average dust concentrations are required—one for inhalation (probably residents are present more during the night than during the day, on average) and one for deposition. Any reasonable model of dust emission, dispersion, and deposition should take account of such factors, because they will include the strong correlation between dust concentration and meteorological conditions. This could be a mistaken identity error, or an extrapolation error.

V_dep is defined as the deposition velocity for dust to soil (DTSC

1998a, p. 792). Its value is given as 500 m/day (CV 0.30) (DTSC 1998a, p. 815). However, the closest parameter in the DTSC documentation (1998a) is "deposition velocity of air particles," which is given as 690 m/day (CV 1.45), suggesting a transcription error. In addition, the measurements used to derive a value in DTSC (1998a, p. 555ff) are not relevant to deposition of wind-blown dust from a landfill. They are (presumably) measurements of ambient particle-deposition rates (DTSC 1998a, p. 560 and 578), and thus correspond to a very different size distribution from wind-blown dust, so that using this parameter is an error of mistaken identity (or possibly an extrapolation error).

It is confusing to introduce a set of parameters and then redefine them using the same names, as occurs for two pathway exposure factors (PEFs) with the introduction of dilution_dust (DTSC 1998a, pp. 791, 792). Dilution_dust should be introduced immediately in the original definitions of the PEFs. As another example, PEF_{inh} in equation 2 (DTSC 1998a, p. 792) does not correspond to the spreadsheet definition of the same term (a documentation error).

Quantitation of Intake for Workers

Section 4.5 of the DTSC report (DTSC 1998a, p. 793) contains many documentation errors. In equation 11, the variables CR_{dust} and CR_{vapor} in the numerator on the right have different dimensions from the other terms, and so equation 11 cannot represent any physical process (i.e., the described model is physically impossible). The body weight (BW) is introduced in the denominator, despite the fact that this is already present in the variables CR_{ing} and CR_{der}. The same sort of errors occur in equation 12, and in addition the variable C_{ing} used in equation 12 is not defined. It is also incorrect to equate the vapor and dust concentrations (below equation 12). Furthermore, the need for Section 4.5 is unclear, because these intakes are never calculated in the spreadsheets, and correctly so (what is required are route-specific intakes for comparison with route-specific toxicity values).

ABS is defined as chemical-specific absorption through the skin (unitless) (DTSC 1998a, p. 794). (It was incorrectly referenced to Table 4; the correct reference is to Table 3 (DTSC 1998a, p. 812)). This factor should not be the absolute absorption through the skin, but an absorption relative to that occurring in the experiment from which the reference dose or cancer potency is defined (these might be different). Because the reference dose and cancer potency are generally expressed in terms of external doses (i.e., the dose crossing the envelope of the human body), they already incorporate the absorption factor applicable during the experiments used to define them. Thus using absolute absorption factors corresponds to an extrapolation error.

DTSC's rationale for changing the algorithms in the PEA model for estimating emission rates for volatile and semivolatile organic compounds is incorrect. Selection of the EPA algorithm (EPA 1995a, cited in DTSC 1998a, p. 795) would require a correlation between the scenario evaluated by EPA (an infinite depth of contaminated soil instantaneously emplaced at time zero, with no infiltration of rainwater) and that to be evaluated by DTSC for the workers on a landfill (small quantities of waste added at intervals over a long time interval, with daily cover applied and possible infiltration of rainwater). The two scenarios are so different as to completely invalidate all the calculations for the DTSC scenario based on the EPA algorithm. DTSC has thus incorporated an inappropriate model. In addition, the waste worker spreadsheet (work_org.xls) implements equation 15 (DTSC 1998a, p. 796) incorrectly. It incorporates an extra factor of 0.1 (in the column headed F/C0) that is undocumented and has no physical basis—an implementation error (although of an inappropriate model).

Furthermore, equating the upper limit (concentration) of the chemical in free moisture in soil to the product of solubility and the moisture volume fraction in soil is incorrect (DTSC 1998a, pp. 795 to 796). What is required is just the solubility (S), as correctly implemented in the spreadsheet (a documentation error). The definitions of bulk_density and part_density (DTSC 1998a, p. 796) are logically inconsistent. One equation is for wet bulk density, the other is for dry bulk density, but that distinction is not made. It must be specified whether bulk_density is wet or dry bulk density (this appears to be a documentation error; the propagation of this error in the implementation has not been checked).

The quantities referred to as concentrations in equation 15 (DTSC 1998a, p. 796) are in fact mass fractions. (Concentrations have dimensions of mass/unit volume; mass fraction is dimensionless.) This common error (some would say convention) appears trivial, but is in fact a documentation error—it leads to confusion when attempting dimensional analyses, and so prevents a useful quality-control check.

The calculation for the apparent diffusion coefficient in soil in equation 16 (DTSC 1998a, p. 797) contains a documentation error. In the numerator of the expression on the right of the equation, the operator connecting the two inner bracketed expressions should be addition, not multiplication. In addition, to correspond with the previous definitions, and with the conventions used throughout the document, there should be a factor of 0.001 liter per cubic meter (L/m^3) multiplying the variable Kd in equation 16. However, this convention should not be followed—there should be no conversion factors present in any of the equations describing the models. Symbols in such equations should represent the quantities involved, including their dimensions, not just their numerical values (which depend on the units in which they are measured).

No adequate reference is provided for the mixing height (MH) above the landfill. In Table 5 (DTSC 1998a, p. 817), mixing height is given as 2 m for a 4.54 × 10^5 m^2 landfill, with a reference to an earlier 1994 DTSC report. Examination of that reference indicates a default mixing height of 2 m for a residential scenario with lot size of 22 m^2, where it is more plausible, although again there is no documentation for selecting this value. For the landfill scenario described here, the value is extremely low. A reasonable estimate for mixing height in this situation takes account of typical plume opening angles. The landfill has length of approximately 670 m, so a 2-m mixing height corresponds to a plume opening angle of about 1/335 radian. If that were correct, practically all the dust raised from the landfill would go underneath the noses of the workers. The committee recommends the use of a model that leads to a reasonable estimate of dispersion over an area source. CalTOX explicitly has such a model and ISC2 and ISC3 both handle area sources; it is inexplicable why DTSC did not use such a model. This appears to be a situation where a parameter value has been taken from one scenario and applied to another, without consideration of the difference between scenarios or the appropriateness of the models used, a type of extrapolation error.

Concentration Limit in Waste

For workers on a landfill (DTSC 1998a, p. 798), the method for calculating a hazard index might be inadequate, because the estimated emission rate and concentration have been averaged over the total exposure time (5.5 years for workers). However, the chosen model for emissions, which is incorrect, corresponds to emission rates that decrease inversely with the square root of time, and so are much higher initially. In the context of its chosen model, therefore, DTSC's averaging method is a structural error. However, as pointed out in Chapter 3 and above, the emission model is incorrect for the scenario evaluated. Emission rates from landfills are likely to vary rapidly and substantially on short (hour to day) timescales as wastes are dumped or surfaces are worked or covered. On longer timescales (days to years), the average emission rates are likely to be much more constant while the landfill is active, but finally decreasing after landfill closure. Such features should be reproduced by appropriate emission models. DTSC should determine whether shorter-term averaging of concentrations for comparison with short-term reference concentrations is required, in addition to the longer-term averages.

Monte Carlo Analysis

The first three sentences of DTSC's discussion of the Monte Carlo analysis (Section 6.2, DTSC 1998a, p. 799) demonstrate a fundamental misunderstanding about probabilistic methods. This misunderstanding has been implemented in the PEA spreadsheets, so that those spreadsheets do not necessarily provide the results that DTSC expects. The problem (Burmaster and Thompson 1995; Burmaster et al. 1995) can be expressed as follows. Suppose a risk estimate R is obtained as a function of various parameters a, b, c, . . . and a concentration C, so that $R = f(a, b, c, \ldots, C)$ for some function f. In a probabilistic calculation, the parameters a, b, c, . . . have probability distributions that induce a probability distribution on R for fixed values of C. This corresponds to the physical situation, and the problem is to find the value of C that ensures that some upper confidence limit on R is equal to some target value.

In a deterministic setting, the equation $R = f(a, b, c, \ldots, C)$ is, in principle, soluble to obtain $C = g(a, b, c \ldots . R)$, allowing a direct calculation of the value of C that corresponds to a given value of R. Unfortunately, while this equation provides an inverse calculation for fixed values, it does not provide the probabilistic inverse.[1] One cannot perform a Monte Carlo procedure with the distributions for a, b, c . . . . and obtain the "distribution" for C given a fixed R—the whole concept of a distribution for C is indeed meaningless. This becomes more obvious when it is realized that, in a probabilistic calculation, the first equation is simply shorthand to indicate that the distribution function for R is a complex convolution integral over all the distribution functions for a, b, c . . .; and the Monte Carlo procedure is just a way to compute that convolution integral.

The PEA spreadsheets have been constructed to estimate a concentration C given a fixed target value for R and fixed values of the parameters a, b, c . . . . When the Monte Carlo procedure is run in one of these spreadsheets, what is obtained is a set of values for C at the fixed target value for R. In general, the lower percentile of this "distribution" for C does not correspond to the concentration that would produce a distribution for R with upper percentile equal to the target value. All of the PEA spreadsheets, in their current form, satisfy the exception to this general rule (see footnote 1). However, Work_org.xls satisfies that exception through a programming error—by design it should not. In Work_org.xls, DTSC (1998a, pp. 795-796) indicates that there is supposed to be a selection of the smaller of two concentrations C_{sat} and C_a[2] to ensure correct calculations when the groundwater and the air within the soil pores are saturated with the chemical. With that test present, the excep-

---

[1]There is at least one important exception to this statement, and that is where R and C are linearly related by a multiplier that is independent of C and R. In that case, the distribution that is calculated is effectively that of the multiplier alone, and probability statements about the upper end of its distribution correspond to similar statements about the complementary probability for the lower end of the distribution of its inverse.

[2]The intent and the physical meaning are clear—to prevent the inappropriate estimation of an evaporation rate using a vapor concentration exceeding that at saturation (of both groundwater and air within the soil pores). This requires a comparison of C_{sat} with C_w, not C_a, another documentation error. The symbol C_a is actually used twice, with different meanings (DTSC 1998a, pp. 795 and 797). The emission model has to be substantially modified if the soil concentration is high enough that the groundwater and soil air are saturated—it is not adequate to simply substitute the saturation concentration into equation 15 (DTSC 1998a, p. 796).

tion noted in footnote 1 does not apply. While the test is necessary to ensure a physically reasonable model, it is actually absent from the spreadsheet. Furthermore, contrary to the DTSC's statement (1998a, p. 795), there are cases where the test is relevant (the groundwater and soil air are saturated at the calculated soil concentration), even with mean values for parameter estimates for methoxychlor and toxaphene, and surely for other chemicals for some parameter values selected during the Monte Carlo procedure. Applying the same procedures to further chemicals (with possibly very different physical characteristics) may also be expected to result in cases where the groundwater and soil air are saturated at the proposed TTLCs.

In contrast, the CalTOX uncertainty analyses appear to be constructed in a way that allows correct calculations (although the committee has not verified that such correct calculations were performed). Provided all the compartment concentrations remain within certain bounds (imposed, for example, by solubility constraints), all the models within CalTOX are linear in the compartment concentrations. Thus exposure point concentrations, and hence risk estimates, are linear functions of the initial compartment concentrations. Given fixed initial compartment concentrations (or similar inputs proportional to such concentrations, like rates of application), a distribution of risk estimates is calculated, and the required percentile of the risk distribution obtained. The initial concentrations (or similar inputs) may then all be scaled by the ratio (target risk value)/(risk at the required percentile of the distribution). The Monte Carlo procedure should then be repeated with the resultant scaled compartment concentrations, to ensure that none of the concentration bounds are exceeded.

The parameter ET_{heavy} (time spent in heavy activity) is documented as having a triangular distribution (DTSC 1998a, p. 800, section 6.2.1), and this is repeated in Table 7 (DTSC 1998a, p. 819). In fact, the work_met.xls and the work_org.xls spreadsheets have a constant value of 3.75 hours with no distribution incorporated (an implementation error).

LeadSpread

DTSC developed a mathematical model called LeadSpread for estimating human blood-lead (Pb) concentrations resulting from contact with lead-contaminated environmental media. A blood-lead concentration of 10 micrograms per deciliter (μg/dL) of whole blood was established as a level that will adequately protect adults and children. The model allows estimation of various percentiles of blood lead concentrations associated with a given set of exposures. The model can be used to establish a concentration in soil or waste that will result in an estimated 95% (or other) upper percentile blood-lead concentration of 10 μg/dL. The model can provide an estimate of blood-lead concentrations resulting from five exposure pathways: dietary intake, drinking water intake, soil and dust ingestion, dust inhalation, and dermal contact.

The DTSC model falls within acceptable standards for models of this type, although the documentation for the model presented in the DTSC report is incomplete. Although a complete list of model parameters is provided, the structure of the model might not be clear to the uninitiated reader.

The general population is exposed to lead in ambient air, foods, drinking water, soil, dust, and fume. The subsets of the general population at highest risk of health effects from lead exposure are preschool-age children, the fetuses of pregnant women, and occupationally-exposed white males between 44 and 59 years of age. Within these groups, relationships have been established between lead exposure and adverse health effects. For the population including children, exposure to lead occurs primarily via the oral route, with some contribution from the inhalation route, whereas occupational exposure is primarily by the inhalation route with some from the oral route.

Blood lead concentration is an integrated measure of internal dose, reflecting total exposure from all sources and over time. Most data relating human health effects to lead exposure are based on blood-lead concentrations, so regulatory standards are typically based on blood-lead concentrations rather than external dose. EPA, the Food and Drug Administration, and the Centers for Disease Control and Prevention have determined that childhood blood-lead concentrations at or above 10 μg/dL may present risks to children's health. As a result, DTSC is following current practice in proposing TTLCs based on a 10th percentile estimate of the concentrations in environmental media corresponding to a blood-lead concentration of 10 μg/dL for children.

Each of the exposure pathways is represented by an equation relating incremental blood-lead increase to a concentration in waste using contact rates and empirically determined ratios. The contribution from each pathway to blood-lead levels is added to arrive at an estimate of median blood-lead concentrations resulting from total exposure. Parameter values used in defining exposure equations fall within accepted ranges; however, one parameter value taken from PEA is known to be incorrect—specifically, the dilution factor in surface soil from the original waste (this error results from a transcription error for the soil-mixing depth in the Off_risk.xls spreadsheet).

CalTOX

The committee has reviewed CalTOX and DTSC's modifications very superficially, because of the short time available. However, as with the PEA model, many of the component models within CalTOX appear to be oversimplifications for the scenarios that DTSC is considering. In addition, the committee found numerous errors, in the context of the DTSC scenarios, in the parameter values used in CalTOX. Some of the specific problems with CalTOX and its associated parameters are described below (these are just examples, not a complete set of all the problems that were found).

For calculations of concentrations of materials in soil, CalTOX uses a three-compartment model, and assumes that the time-variation of all concentrations can be adequately modeled by linear first-order differential equations with constant coefficients. Such a model does not correspond to the physical processes, however, at least in some situations in DTSC's scenarios. This mismatch was explicitly realized by the designers of CalTOX, and the model parameters are specifically chosen to match a more physically realistic mathematical model, the Jury model, in order to take into account some of the mismatch (DTSC 1998a, pp. 263–268). The committee is concerned, however, about the adequacy of such matching for DTSC's purposes. It questions whether such approximations are either necessary or desirable and whether DTSC took into account the specific limitations on the model adjustments noted in the CalTOX manuals, and apparent in the simulations described there (DTSC 1998a, pp. 265-267).

As specific examples, the CalTOX parameter value estimates were matched to minimize differences in the surface fluxes and the soil-mass inventories in two of the CalTOX compartments; but the potential resulting three-fold uncertainty in even these quantities is not incorporated in DTSC's calculations. Moreover, although such matching might ensure that some results are reasonably accurate for some pathways (e.g., long-term estimates for inhalation) in some conditions, the same does not necessarily hold for those pathways under other conditions, or for other pathways of importance. For example,

- Short-term (up to approximately a year) emission rates of vapors might be substantially higher than predicted by CalTOX, resulting in substantial underestimates of short-term exposures that might be toxicologically significant (DTSC 1998a, p. 268).
- Soil concentrations in the top few millimeters of soil, most relevant for the dermal contact pathway, are substantially overestimated by the CalTOX approach (DTSC 1998a, pp. 266-267).

The landfill gas model introduced by DTSC during the modification of CalTOX (DTSC 1998a, pp. 78–104) is also open to question. This model assumes a physically unrealistic continuous flow of landfill gas at a constant rate that is time-independent; it is also physically unrealistic because the variation of flow rate with the mass of material in the landfill has a discontinuity at 1.1 million tons. The committee wonders why the time-dependent EPA landfill gas generation model, for example, was not used. The manual describing this model is available at http://www.epa.gov/ordntrnt/ORD/WebPubs/landfill/index.html.

The values and uncertainties of the various parameters used in CalTOX and the other models come from "The Distribution of California Landscape Variables for CalTOX, February 1996" (DTSC 1998a, pp. 555 ff.), similar reports for individual chemicals (e.g., Hsieh et al. 1994), and other sources such as those listed as references in DATCAL.XLS and the accompanying DATREF.XLS. However, the values and uncertainties or variabilities reported in these sources occasionally do not correspond to those required; and there appear to be multiple transcription errors. A few examples follow. (Chapter 3 also has a short discussion of some pervasive extrapolation errors in the estimation techniques used by DTSC to estimate some of the parameter values used in CalTOX, for example, partition coefficients for plant tissue and biotransfer factors to meat, milk, and eggs.)

Dust-Deposition Velocity

The reported dust-deposition velocity (DTSC 1998a, p. 555) corresponds

ISSUES OF MODEL APPLICATION 101

to the mean and standard deviation of many individual measurements of deposition velocity, not to the required long-term average velocity at any particular location and the variability between locations. The deposition velocity at any one location will vary substantially from time to time, but that is irrelevant for the modeling required by the DTSC scenario, an error of mistaken derivation. Moreover, there is also a transcription error: 500 m/day (CV 0.30) (DTSC 1998a, p. 816, table 5) versus 690 m/day (CV 1.45) (DTSC 1998a, pp. 560, 578).

Organic Carbon Content of Residential Soil

The reported mean value for the carbon fraction of residential soil appears to be extremely low for fertile soil, but may be biased by many low measurements in infertile, but residential, soils, as the range is up to a reasonable value. However, for the CalTOX modeling in which this parameter is used, the range of values for fertile soil (backyard gardening pathway) is required, not an average over all residences. This might be an error of mistaken identity or derivation. The reported mean of the carbon content of residential soil is 0.003 (CV 0.367) (DTSC 1998a, pp 560, 572); but in Table 5, it is given as mean 0.003 kg/kg (CV 3.67) (DTSC 1998a, p. 815) (this error occurs twice, in Table 5, Section 2 and Section 5) and thus has been incorrectly transcribed.

Molecular Weight

DTSC gives the molecular weights of all individual chemicals as a mean and CV, based on several values found in the literature. For example, the molecular weight of trichloroethylene is given as 131.4, with a CV of 0.00039 based on five values (DTSC 1998a, p. 736). Inspection of the original data sources indicates that one value used was simply in error, and the other four values represent the same value written with differing numbers of significant digits. The rote approach taken by DTSC to evaluate a "distribution" for molecular weight undermines the committee's confidence in DTSC's selection of any parameter and of the modeling effort in general. The uncertainty in molecular weight for these chemicals of know structure is negligible, and the DTSC approach does not address the remaining uncertainty.

Chemical Properties

Many chemical properties are given in Table 2 (DTSC 1998a, p. 808) with a citation to the DATCAL.XLS spreadsheet. The committee has a copy of this file and the associated DATREF.XLS file. Although both files contain some references, in many cases there is no indication as to the original data source from which these values were taken. For example, there is no explanation or reference as to why the CVs for the vapor pressures for chlordane and TCDD are equal to 1.58 and 1.57, respectively, whereas many CVs for other substances are much smaller.

ANALYTICAL METHODS

California is proposing to no longer require the use of the waste extraction test (WET) for determining the extractable constituents of hazardous wastes not classified under the Resource Conservation and Recovery Act (RCRA), relying rather on the use of EPA's toxic characteristic leaching procedure (TCLP). The TCLP has long been required by EPA to define the toxic constituents of RCRA hazardous wastes.

In 1972, California's then-new Hazardous Waste Control Act defined "hazardous waste" and "extremely hazardous waste" and, in 1977, California added the requirement that the state develop and adopt criteria and guidelines for the identification of these two waste categories. The California Assessment Manual (CAM) (California Department of Health Services, 1981) prescribed the use of WET as the state's test procedure. The CAM-WET test extracted solid wastes with pH 5 citrate buffer for 48 hr (Table 4-2).

At the federal level, the 1984 amendments to RCRA led to adoption by EPA of a batch extraction test, called the extraction procedure, which was designed to simulate processes occurring in landfills and that might

TABLE 4-2 Comparison of Conditions for WET and TCLP

Test Conditions	WET	TCLP
Solid-to-Solution Ratio	1:10	1:20
Buffer	Citrate, pH 5	Acetate, pH 5
Time	48 hrs	18 hrs
Enclosure Status	Not enclosed	Closed system with zero headspace

Source: DTSC (1998a, p. 1114).

contribute to the leaching of toxic constituents. EPA subsequently replaced the extraction procedure with the TCLP (55 Fed. Regist. 11798, March 29, 1990). In the standard version of the TCLP, a pH 5 acetate buffer is used in a 18-hr extraction test (Table 4-2).

A comparison of WET and TCLP, using California wastes and waste composites, was provided by DTSC in the Regulatory Structure Update Extraction Test Project Summary Report contained in the DTSC report (DTSC 1998a, p 1078). WET consistently extracted more of 10 elements than TCLP (Table 4-3), with the exception of one mercury result. For several waste-element combinations, WET extract concentrations exceeded TCLP extract concentrations by 1 to 2 orders of magnitude. The major difference in the two extraction procedures is that the citrate buffer in WET leads to chelation of some elements (e.g., lead), and the direct release of elements bound in the solid-phase by dissolving high content metals (e.g., iron) by chelation (DTSC, 1998a). Although WET extracted more of the test elements than TCLP, a comparison of results with municipal solid waste leachate (MSWL) indicated that WET is generally more exhaustive than TCLP, leading to significant overprediction of what is actually present in the leachate for many elements (Table 4-3). On balance, TCLP gave a better representation of what actually leaches from these landfills for most, if not all, elements. Thus WET generally overestimates what leaches out of landfill waste over the lifetime and post-closure period of a landfill, whereas TCLP's results in leaching simulation are more in line with observed leaching behavior. In subsequent tests, citrate, which is used as a buffer in WET but not in TCLP, has not been found to be a constituent of leachate from California landfills. For these reasons, and for the sake of harmonizing with EPA by requiring that only one test in California, DTSC has proposed replacing WET in favor of TCLP for its non-RCRA solid hazardous waste classification testing program.

For an exact simulation of landfill leachates, neither WET nor TCLP provides satisfactory performance for oily wastes, for volatiles that might reach groundwater by diffusion, or for some elements occurring as oxyanions, such as arsenic, chromium, molybdenum, and selenium. Also, neither test adequately addresses questions of speciation for chemicals that can exist in more than one form, such as element, salt, and anion. WET overestimates the leaching potential for many elements in representative California landfill wastes, but there are several exceptions to this, such as cadmium, nickel, and thallium.

Based on the shortcomings of both the WET and the TCLP, and the

TABLE 4-3 Comparison of WET and TCLP in Short-Term Extractions (mg/L)[a]

Substance	WET	TCLP	Max MSWL
Arsenic	6.51	0.13	2.08
	49	0.06	2.07
	4.9	0.10	0.13
Beryllium	0.02	<0.001	0.00
	0.02	<0.001	0.01
	<0.01	<0.001	0.01
Cadmium	23	11.85	27.7
Cobalt	0.87	0.42	0.87
	<0.20	0.02	0.03
	0.83	0.07	0.02
Mercury	0.02	0.575	0.19
	0.03	0.003	0.01
Molybdenum	1.27	<0.030	0.45
	<0.3	<0.030	0.44
	0.84	<0.030	0.04
Nickel	174	163	334
Lead	391	11.1	19.1
	275	11.9	5.05
	16.80	1.750	1.80
Selenium	<0.80	<0.080	1.43
Thallium	3.79	1.500	4.45

[a] Data are for municipal solid waste landfill leachates (MSWL) from Hyperion (Los Angeles), Los Gatos (Guadalupe), Lodi, and Ukiah, California.
Source: Adapted from Table 7 (DTSC 1998a, p. 1060).

fact that both test procedures have beneficial features (exhaustive extractability of WET, simulation more reflective of actual leachate content and acceptability of TCLP), the committee supports the development of a single test protocol to classify California's hazardous waste, and to do so in harmony with the classification test of EPA. Such a test should provide results that can be related to field-realistic exposures, including the uncertainties associated with leaching pathways in the field. Understandably, DTSC may choose not to pursue this effort alone based upon limitations in needed resources. Such an effort would also be quite time-consuming because of its nationwide implications and the need for extensive testing and validation under a variety of waste, climatic, and soil conditions. The committee recognizes that the TCLP has nationwide status, use, and acceptability. Harmonizing California's extraction test with that required by EPA would minimize the testing burden on waste disposers in California, who would need to conduct only the TCLP. However, DTSC has not yet provided convincing arguments, either to the committee or, based on written comments, to stakeholders, for the sole adoption of TCLP and elimination of WET. The committee recommends that DTSC conduct an open evaluation of the experimental evidence, including the results of side-by-side testing and the opinions of its own staff, federal EPA counterparts, and stakeholders, before reaching a conclusion on the three possibilities before it: (1) adopt the TCLP as the sole test; (2) continue requiring both the TCLP and the WET; or (3) develop a new test that overcomes the deficiencies of both the TCLP and the WET.

There is one very important aspect of the use of either WET or TCLP assays that DTSC has overlooked in its modeling effort and that it should bear in mind in further evaluations. DTSC is currently using these assays as though they exactly match field conditions; indeed, much of the argument and experimental program has gone into the evaluation of how well each of them matches field conditions. However, a probabilistic approach needs to explicitly introduce the uncertainties in extrapolations such as those from a laboratory assay to the field, and this DTSC has failed to do in considering either the WET or the TCLP methods. Either assay could be used in a probabilistic procedure, although each would have different uncertainties associated with its use. DTSC has expended much effort in a commendable experimental evaluation of WET and TCLP, and the experimental results appear to provide a suitable basis for evaluating the uncertainties associated with the results of those assays with leaching in field conditions. The different biases and/or larger uncertainties associated with certain types of chemicals, or certain types of waste stream, can be built into the probabilistic modeling.

For the (DTSC-designated) category 2 elements of arsenic, antimony, molybdenum, selenium, and vanadium, DTSC proposes to use unadjusted TTLCs for arsenic, molybdenum, and antimony, and to use either the detection limit or develop a new test for selenium and vanadium. Use of the detection limit has the disadvantage of being driven by the state of analytical methodology rather than risk, contrary to the aim of the DTSC program. Also, it is somewhat arbitrary, for example, in its use of the analytical limit of detection (LOD) rather than analytical limit of quantitation (LOQ) or twice the LOQ. Similarly, DTSC proposes to use twice the estimated quantitation level (2X EQL) in lieu of a SERT when the calculated concentration of the SERT is less than the EQL. In both cases, the committee emphasizes that there is no connection between the sensitivity of chemical analytical methods and the sensitivity of biological receptors, thus, the use of 2X EQL to establish a SERT is also not risk-based.

Analytical methods are continually being improved as new instrumental and other techniques are introduced, and detection limits vary from laboratory to laboratory and sample to sample. Detection limits, and limits of quantification, may be influenced by background. This needs to be taken into account when analyzing for naturally-occurring substances (mercury, selenium, cadmium, etc.), which may vary in background concentration from location to location. It also needs to be taken into account for organic contaminants for which the matrix may contain substances that mimic or interfere with the analyte of interest. This matrix effect may also vary from sample to sample and location to location. Biological sensitivity is fixed by the inherent toxicity of the analytes and response of the organism being exposed to the analyte under specified conditions.

Comparative testing to determine if the use of detection limits as proxy values is protective under reasonable exposure scenarios is lacking. It might turn out that such proxy values are protective, but this can not be determined from the information provided to the committee. DTSC should undertake, for example, a comparison of the SERT values with the EQLs. This should be done by evaluating a range of compounds with different toxic potencies and EQL values to determine the degree of protectiveness.

TOXICITY TESTS

Tests Related to Human Health

The proper evaluation of the potential adverse health effects of a substance requires knowledge of the chemical and physical properties of the test material; anticipated human exposure conditions, including environmental levels, duration, pathway(s) and populations; the nature of the anticipated acute and immediate effects or delayed or chronic effects; and (usually) at least one appropriate nonhuman (e.g., animal) model. Only acute toxicity tests will be addressed in this section.

The interaction between the assessment of risk from acute toxicity and the assessment of risk from chronic toxicity is not entirely clear from the flow chart in the DTSC documentation (DTSC 1998a, p. 36). From the figure, one would assume that the first screen is for chronic toxicity as assessed by the development of TTLCs and SERTs, followed by assessment of risk from the acute toxicity of the chemicals based on acute toxicity assays. However, only 38 chemicals have passed through the first (TTLC) screen and, as noted in previous chapters, there does not appear to be a clearly defined method to either add or delete chemicals from either the TTLC or SERT lists. The chronic toxicity risk assessments are based on reference doses or concentrations, or cancer potency factors that were designed to protect the general population, including sensitive subpopulations. Thus, the thresholds developed based on chronic low-dose exposures of the general population would be expected to be much lower than the thresholds that might be developed for acute toxicity based on almost any acute exposure scenario.

The acute oral toxicity thresholds are based on doses or concentrations calculated to be lethal for half of the test animals (LD_{50} or LC_{50} values), divided by a safety factor of 100 and multiplied by an estimated ingestion "rate". The rate given is 5 mg/kg of body weight for children. This is not a rate but rather a dose, although the value is said to be derived from a percentile of the CalTOX parameter corresponding to a rate (the soil ingestion rate). A rate would be a dose per day or some other unit of time. Because only a dose is given, it appears that the threshold is designed to protect someone who, in a one-time, or at least infrequent, situation, actually eats the waste directly, but does not eat it on a daily or regular basis. The acute toxicity threshold so derived could be considered to be protective against lethality for such a one-time ingestion event, but would not necessarily be protective against more subtle toxicity, particularly if the ingestion occurred on a repeated basis. DTSC needs to clarify the purpose of the acute toxicity thresholds, who is to be protected by these thresholds, and whether it expects the exposures to occur one-time or be repeated.

For the acute dermal toxicity thresholds, DTSC provides a better description of the parameter values used in the derivation of the thresholds and correctly uses a dermal contact rate (in millgrams per kilogram per day). A minor point is that the DTSC text refers to oral LD_{50} values rather than dermal values (DTSC 1998a, p. 72).

DTSC presents the use of the acute oral and dermal toxicity thresholds as though they are based on various acute exposure scenarios (DTSC 1998a, pp. 72-74). This is a reasonable approach, but DTSC has not presented clearly defined goals and appropriate scenarios to meet those goals. For example, is it DTSC's intent to protect the most sensitive subpopulation from death if, in a one-time situation, a member of that population wanders on site and eats the waste? Or is it to protect a person who occasionally wanders on the site and eats the waste once a week? Or is it to protect those who live near the waste site and might inhale vapors and particles emitted from the site on a daily basis? The use of oral and dermal LD_{50} values is apparently for scenarios in which there is a high-end ingestion or dermal contact with raw waste streams by a child. The parameters selected for oral ingestion and dermal contact rates fail in this purpose, however, through an error of mistaken identity. What are used are upper percentiles of ingestion and dermal contact rate parameters derived for use in CalTOX; but the distributions of those rates for CalTOX should correspond to the variabilities between individuals in long-term average rates. What are required for the acute scenarios are distributions that also include day-to-day variability for individuals. However, because the scenarios are not adequately described, it is not clear with what frequency the estimated dose will be consumed.

For the acute inhalation thresholds, no exposure scenario is presented, merely a rationale that corresponds to a highly unlikely, and maybe impossible, situation. For vapors, the assumption appears to be that persons could be exposed to vapors in equilibrium with fresh waste undepleted by off-gassing (the committee assumes that the temperature of 250°C specified on page 73 is a misprint for 25°C). Although a scenario for a waste worker might be constructed in which such a situation is possible, it is doubtful that there are any such situations involving the general public; moreover, workers should be protected at lower levels by Occupational Safety and Health Administration (OSHA) standards. The basis for the rationale for the particulate inhalation thresholds is even less secure. What scenarios can DTSC suggest that would result in acute exposures that are limited to the OSHA time-weighted average standards, or the long-term National Ambient Air Quality Standard for particulate matter?

The committee found incorporation of different safety or uncertainty factors for the different acute thresholds also to be questionable. For the oral and dermal exposures, a safety or uncertainty factor of 100 is incorporated; for particles, a safety factor of 10 is included, and for vapor exposures, no safety factor is proposed at all. As for the TTLC and SERT derivations, a consistent approach requires DTSC to make explicit its protection goals, and then to evaluate scenarios with parameter values that correspond to those goals.

It is not appropriate to use only acute toxicity tests and short-term exposure scenarios, rather than chronic toxicity or long-term exposures, as the basis of risk assessments for waste classifcation and disposal. Both types of information have an important place. For chemicals for which there are no TTLC or SERT values, the risk assessments should not be based solely on acute toxicity, that is, by using bioassays with the crudest of endpoints, lethality if chronic toxicity data are available. Such an approach would not give any consideration to reproductive and developmental toxicity, any chronic toxicity (including cancer) or genetic toxicity. At the very least, DTSC should review readily available chronic or other effects data (genetic toxicity, reproductive and/or developmental toxicity) for each of the waste components and compare the concentrations of the components in the waste with the concentrations found to cause no, low, or infrequent effects. Possible sources of such chronic effects information include EPA's maximum contaminant levels for drinking water, oral reference doses or inhalation reference concentrations, or cancer potency factors, and the Agency for Toxic Substances and Disease Registry's minimum risk levels. These values are easily accessed for numerous chemicals and most have been subject to scientific peer review.

If only acute toxicity data are available for the risk assessment, DTSC should follow standard practice and use an additional uncertainty factor to account for the lack of data regarding potential chronic toxicity at concentrations that are lower than those causing acute toxicity. DTSC should take into account the slope of the dose-response curve for acute toxicity data when choosing the uncertainty factor, if such data are available. Failure to use an appropriate uncertainty factor my seriously underestimate the risks associated with chemicals for which only acute toxicity data are available and may result in unprotective thresholds. If the waste contains several chemicals for which chronic toxicity data are available (e.g., several polycyclic aromatic hydrocarbons, which act at common sites to exert their toxicity) then the additive, synergistic, or antagonistic effects of these chemicals, if known, should also be considered when assessing the risk posed by the waste.

It appears that if a waste does not contain any of the TTLC or SERT chemicals, and is classified as nonhazardous on the basis of its acute toxicity, it is not subject to further scrutiny but may be disposed in nonhazardous waste landfills or by other methods such as recycling or land application. As a result, a waste that may pose serious chronic or mutagenic risks at concentrations far below those that cause acute effects and where long-term exposure may be expected as a result of its disposal, may be inappropriately classified as nonhazardous using DTSC's current or proposed classification system. In essence, the use of acute effects data permits higher (less conservative) risk thresholds for wastes than would be possible if chronic effects data for chemicals without TTLCs or SERTs were required. It appears to the committee that the current and proposed DTSC methods provide distinct disincentives for the identification of chronic effects data for particular wastes or waste constituents, since any such identification is likely to result more stringent regulation.

DTSC should also consider the inclusion of respiratory, ocular, and dermal irritation testing as well as allergic sensitization testing, in its battery of acute toxicity tests. The nuisance factor of odors may also have to be taken into account to meet some goals. Members of a community living close to a waste site are more likely to be aware of and concerned about acute effects related to the irritant and odor properties of the waste than any other type of toxicity. Respiratory irritation might exacerbate existing health conditions such as asthma. If more than a single short-term exposure is anticipated (e.g., in waste workers or those living near a waste-disposal site), the potential for sensitization (allergenicity) may be relevant.

A further problem related to the acute and chronic effects of specific chemicals is that the DTSC approach does not take into account the speciation or chemical form of metal contaminants. This is an arbitrary simplification that is not based on true risks. For example, chromium (III) at low doses is an essential nutrient for humans, whereas chronic exposure to chromium (VI) has been associated with lung cancer in humans; the toxic effects of elemental chromium are relatively unknown. Some consideration of the species and chemical form of the metal contaminants present should be attempted for both acute and chronic risk assessments.

Tests Related to Ecology

DTSC proposes to protect aquatic organisms by classifying wastes using acute lethality to fish. Two thresholds based on acute lethality (96-hour LC_{50}) to fish of extracts are used to establish the category to which a waste will be assigned. The first threshold, at an LC_{50} of 30 mg/L, is used to classify a waste as hazardous, and the second threshold, at an LC_{50} of 500 mg/L, is used to distinguish between nonhazardous and special wastes. The 30-mg/L value is derived from 500 mg/L divided by 18, the 10th percentile estimate for the liner protection factor. The current threshold for classifying waste as hazardous is based on a 96-hour LC_{50} value of 500 mg/L (22 California Code of Regulations § 66261.24 (a)(6)). DTSC proposes to retain this regulation but to use only a fish acute lethality bioassay to bring wastes into the lower tier of hazardous waste. It appears that even if a waste was not classified as hazardous based on comparing its concentration with a TTLC, it could still be classified as hazardous based on the results of the fish acute lethality test. It is unclear from the DTSC document if or how SERTs will be applied in the classification of wastes in the ecological scenario. It would appear that for wildlife, total concentrations of chemicals in wastes will be compared with TTLCs and a fish acute lethality test will be performed. The fish acute lethality test does not include the potential for bioaccumulation or biomagnification and would not be useful for compounds that are chronically toxic and have great acute to chronic ratios.

The proposed methodology assumes that fish are the most sensitive aquatic organisms. This is certainly not always the case, for some aquatic organisms are more sensitive than fish to a number of compounds. Thus, the proposed screening methodology might not sufficiently protect aquatic life or wildlife that eat aquatic organisms. Also, DTSC does not specify how a waste would fail the bioassay test. Presumably, if the TCLP leachate causes greater than 50% lethality of the fish, the waste will be classified as hazardous. The committee concludes that the use of an acute bioassay using fish would not be sufficient to protect aquatic organisms or animals that might eat aquatic organisms.

Retaining consideration of aquatic toxicity in the screening system is appropriate and is supported by the committee; however, the selection of a threshold value of an LC_{50} for TTLC for listing a material as hazardous is considered to be somewhat arbitrary and has no scientific justification. This is not a risk-based approach. For the approach to be risk-based, DTSC must consider exposure and dose or concentration simultaneously when establishing a risk threshold. The risk presented by a waste is a function of exposure concentration and a threshold for acute effects; thus, setting a single value for a threshold is inappropriate. Although a single value might be predictive, it has not been demonstrated that it will be protective relative to possible aquatic concentrations.

5

Meeting Program Goals

IN THIS CHAPTER, the NRC committee identifies broader programmatic issues that, if addressed, would enhance the risk-based approach of the California Environmental Agency Department of Toxic Substances Control (DTSC). As described in Chapter 1, DTSC undertook revision of their current waste-classification program to respond to the aims of the California regulatory structure update (RSU). Of the four guiding principles of the RSU, two bear on the scientific or technical aspects of the programs (protecting public health and the environment and fostering compliance through regulatory flexibility and simplicity), and the other two are more policy-based and were not considered by the committee (remove regulations that are ineffective and coordinate regulations with other governmental agencies). The first two principles and DTSC's approach to them by using a risk-based waste-classification scheme are discussed below, along with suggestions for evaluating the success of the new waste-classification regulations.

This chapter then examines whether DTSC's proposed waste-classification scheme would likely achieve the scientific and technological objectives propounded by the DTSC, which are to (1) consider exposure when classifying wastes; (2) develop a mechanism for incorporating new toxicological or technical information; (3) provide guidance on handling specific waste streams rather than specific chemicals; and (4) institute a system for regulating chemicals other than the current 38 total threshold limit concentrations (TTLC) chemicals or the 36 current soluble or extractable thresholds (SERT) chemicals.

Finally, there is a discussion of some other concerns DTSC may want to consider for improving their waste classification system. These include integration of SERTs and TTLCs into a single risk-based value, development of a true multimedia risk-assessment approach for ecological end points, and expansion of the universe of risks that are considered when classifying wastes.

RSU GUIDING PRINCIPLES

Protect Public Health and Environment

The committee strongly endorses DTSC's proposed use of a risk-based, rather than a simply toxicity-based, waste-classification system. An integral aspect of the proposed DTSC system is a change from a single toxicity-based threshold to distinguish hazardous and nonhazardous wastes to a risk-based system with two thresholds, one to separate nonhazardous from special wastes and a second more conservative threshold to separate special wastes from hazardous wastes. Special wastes will be subject to less stringent disposal and reporting requirements.

Although the current waste-classification system is far from perfect, there is no indication from DTSC or public commenters that it is not protective of the public health and the environment, at least for the 38 chemicals that are included in the system. The proposed classification, with two risk-based thresholds, will increase the number of wastes that may be classified as special. According to DTSC's proposal, wastes currently classified as hazardous may be reclassified as special and be disposed to landfills that are less restrictive. Wastes that are currently classified as nonhazardous may also be reclassified as special and subject to more stringent disposal requirements. Although DTSC indicated at the public meetings that the new waste-classification system will be more protective of human health and the environment, the documentation fails to support this contention. Specifically, it is not demonstrated in the DTSC documentation that the broader special-wastes category will protect human health at the specified risk level. Indeed, for silver, the proposed system is theoretically less protective by removing silver from the TTLC list completely. The inclusion of several examples showing how different wastes would be classified (and subsequently disposed) under the current and the proposed systems might go some way to demonstrating DTSC's contention that the proposed approach is at least as protective as the current system, if not more so.

DTSC has not fully documented the limitations and safeguards for the disposal of special wastes in class II landfills. The committee is unable to determine whether such disposal is appropriate for the necessary level of health and environmental protection. Documentation for the classes of landfills or other waste management options, such as land conversion, is an essential part of an integrated risk assessment.

There are two issues in the proposed system concerning ecological risks that need to be highlighted. In the first case, the proposed approach states that lower TTLC values, based on the scenarios and parameter values selected to be appropriate to protect humans, will necessarily be protective of wildlife. There are a number of examples in which this is not the case, such as DDT. It is incumbent upon DTSC to provide an analysis that gives the probability that the exit-level TTLCs derived for the protection of human health would be protective of ecological, nonhuman receptors.

The second issue concerning ecological risks is identical to that of using a two-tiered system for human health. Does a two-tiered system have any advantage for environmental protection? The proposed system also uses an upper and lower acute toxicity threshold based on an acute fish toxicity test for the upper threshold (nonhazardous versus special wastes). This toxicity value is then divided by a liner protection factor to distinguish between special and hazardous wastes. The advantage of using this system over one with a single threshold is not specified. Similarly, the advantages of a two-tiered system for protecting human health and the environment are not specified in, nor are they obvious from, the DTSC documentation. This suggests that the more intrinsic goal of the two-tiered system is to increase regulatory flexibility and simplify the regulatory requirements.

Regulatory Flexibility and Simplicity

A second principle of the RSU is to increase regulatory flexibility and simplify regulations. DTSC intends to accomplish this by increasing the number of wastes that will be categorized as special based on the two risk thresholds. The advantages of having a waste designated as special is not obvious from DTSC's documentation; the waste classification system given in the report (DTSC 1998a, p. 36) suggests that the new system is more complex than with the one-tiered system.

Given the proposed exposure scenarios, and in particular their lack of definition of protection goals and appropriate model parameters, it is possible that a significant number of compounds will move from the nonhazardous to the special-waste categories. At the public meetings, DTSC indicated that the volume of hazardous and nonhazardous wastes will not change significantly under the new proposal (DTSC, personal commun., September 10, 1998 and November 20, 1998); however, they have not presented a convincing argument supporting this statement. DTSC should reevaluate the flexibility created by the addition of the special-waste category after it has better defined the populations to be protected, protection goals, exposure scenarios, and model parameters consistent with those goals.

The proposed system appears to lack flexibility with regard to the addition or deletion of chemicals. As discussed below, there is no clearly defined mechanism for selecting new chemicals for SERT or TTLC development or for removing chemicals from the SERT or TTLC lists should there be new evidence that they pose little or no risk. This might also be a consideration for chemicals already on the list if speciation is an important factor in the risk posed by the chemical. For example, metals are regulated as elemental forms with no distinction between inorganic or organic forms (with the exception of lead). These forms might have significant differences in toxicity and subsequently risk. Chromium is a good example for which toxicity varies considerably with the speciation of the metal.

DTSC PROGRAM GOALS

DTSC indicated that it had several goals when revising the current waste-classification system. California has determined that the federal Resource Conservation and Recovery Act (RCRA) classification system was not sufficiently protective of human health or the environment and that it was necessary for the state to identify and manage those wastes that might not be classified as hazardous under RCRA but that might pose risks to Californians. The current waste-classification system, as summarized in Chapter 1, is limited. Specifically, DTSC wished to improve the system by doing the following:

- Considering exposure in classification of waste.
- Incorporating new toxicological or technical data.
- Handling specific waste streams rather than specific chemicals.
- Developing a mechanism for regulating chemicals other than the 36 SERT and 38 TTLC chemicals.

Considering Exposure in Classification of Waste

DTSC has prepared a good prototype system for incorporating risk into the classification of waste. Using models, such as CalTOX, it has factored exposure in with chemical hazards to develop risk-based thresholds for selected waste components. This has been done for human health risks from exposure via groundwater, surface water, air, soil, and food exposure pathways. Incorporating exposure data including information on the environmental fate and transport of chemicals, as well as the identification of potential receptors (human and ecological), can only increase public confidence that the proposed regulations will protect their health and the environment. Risk-based thresholds for human health have been developed for 36 chemicals for groundwater exposure (SERTs) and 38 chemicals for other exposure pathways (TTLCs).

Exposure pathways, however, have not been included in the development of ecological thresholds. The aquatic toxicity values are toxicity-based with little or no consideration of the various pathways by which numerous receptors might be exposed to chemicals. Exposure models are available for ecological risk assessments and their use would significantly improve DTSC's waste-classification system.

However, DTSC has not been clear as to what human or ecological populations are being protected and what level of protection is required or acceptable. Without such specification, it is difficult to develop the appropriate exposure scenarios and to choose the appropriate parameter values for the exposure models.

Incorporating New Toxicological or Technical Data

The incorporation of new chemical or model-specific information into the regulatory process is an important goal for any regulation designed to protect human health and the environment, and is one of the primary aims of DTSC's proposed system. The documents provided to the committee do not describe the mechanism by which DTSC will collect, review, and update such information in the future. No system was presented for periodic or episodic review of SERTs or TTLCs should new toxicity or exposure information become available. For example, if the U.S. Environmental Protection Agency (EPA) were to change the maximum concentration limits on which a TTLC or SERT was based, how would DTSC account for this change in value? Would the new information be considered on an "as available basis," or would there be a systematic review similar to that provided for reviewing the minimum risk levels developed by the Agency for Toxic Substances and Disease Registry for chemicals found at Superfund sites?

As presented in the DTSC report, the derivation of risk-based values would appear to be a one-time event. The development of the upper and lower TTLCs and SERTs are based on an analysis of recently available toxicological information, including reference doses and maximum contaminant levels. Reference doses and maximum contaminant level values are developed by the EPA and are subject to periodic revisions as new toxicological data become available. Supporting documents, which present the toxicological database upon which the health-effects values are based, are also developed for each chemical.

In tandem with the need to update TTLCs and SERTs with new toxicological information is the need to incorporate new exposure assessments. The exposure assessments are based on exposure models such as CalTOX and the preliminary endangerment assessment model. Exposure models are subject to multiple iterations, being updated and refined as new fate, transport, and other exposure information becomes available. The exposure scenarios currently being used could subsequently be found to be inappropriate and would then need to be modified to reflect real-world situations. DTSC has not proposed any method to identify and collect new exposure data based on further environmental monitoring or testing, or additional modeling.

Developing a Mechanism for Regulating Chemicals Other Than the 36 SERT and 38 TTLC Chemicals

DTSC indicated that its new approach will be useful for regulating new chemicals and wastes in addition to those for which TTLCs and SERTs have already been developed. Essentially any chemical might be considered for TTLC or SERT development; however, DTSC does not indicate whether there will be an attempt to select chemicals with potentially greater risk because of greater toxicity, greater exposure, or other criteria. This is an important consideration and has been used with ranking and screening criteria development in systems such as EPA's toxics release inventory and the Occupational Safety and Health Administration's permissible exposure limits. Selecting chemicals for TTLC or SERT development might require the concurrent development of screening mechanisms and ranking processes to identify those chemicals found in wastes that pose the greatest risks. Indeed, this is one criticism of the proposed list of TTLC and SERT chemicals. Many of the TTLC and SERT chemicals are no longer in commerce and will occur in wastes primarily from dredging or recycling of contaminated media, not from industrial wastes. Those chemicals that are highly toxic or persistent in the environment should be distinguished from those known to be in commerce or commonly found in waste.

DTSC should develop a process for selecting chemicals to be considered for SERT and TTLC development. This process should be based on population and environmental protection goals outlined at the beginning of the waste-classification process.

OTHER CONSIDERATIONS FOR DTSC's APPROACH

Other areas where DTSC might expand the utility of its proposed waste-classification system or where the system might be improved are briefly discussed below.

Integration of TTLCs and SERTs to give just one value protective of human health and the environment via groundwater or other possible exposure pathways.

The distinction between TTLCs and SERTs is not entirely justified. In a true multimedia, multipathway approach, only one risk level should be developed for each chemical, any new chemicals, and preferably, for entire wastes. In real-life situations, people and other organisms might be exposed to chemicals in the groundwater through ingestion of groundwater as drinking water. However, they might also be exposed to chemicals from the same source by inhalation of vapors evaporating from such water, and by these and other routes through entirely different pathways from the same source. Therefore, the committee recommends that the exposure modeling include all media and all exposure pathways, and not separate out groundwater exposure. For example, the CalTOX multimedia model is fully capable of including exposure assessment for groundwater along with soil and air. A single regulatory value for all exposure pathways will simplify the waste-classification system.

Multimedia risk assessment for ecological impacts.

Ecological risk assessments are subject to the same multimedia and multipathway considerations as are human risk assessments. However, DTSC's proposed approach is seriously lacking ecological exposure scenarios and subsequent risk assessment values. Use of a toxicity value such as an LC_{50} for fish or EPA's ambient water quality criteria does not constitute an ecological risk-assessment with some exposure assessment and, therefore, should not be called such in the DTSC documentation. Fish are not always the most sensitive species nor is surface water always the primary exposure pathway. As has been demonstrated in the past with numerous pesticides, birds are frequently at a greater risk from exposure to soil pathways, and thus might be the most affected species. Given the exposure scenarios presented in the DTSC report, particularly the land conversion scenario, the presence of contaminants in soil and subsequent ingestion by birds or other wildlife that consume soil (inadvertently or otherwise) might be of significant concern. Therefore, DTSC is encouraged to be more explicit and comprehensive in developing its ecological risk assessments. As presented, ecological concerns are not paramount, and it is impossible to ascertain whether wildlife will be adequately protected under this waste-classification approach.

Expand consideration of potential risks.

DTSC's proposed approach is a risk-based waste-classification system that results in a change in how some wastes are managed after generation, and possibly, in how they are disposed. The committee commends DTSC's use of risk-based analyses, and the integration of these analyses into waste regulations that protect human health and the environment. However, it might be advantageous for DTSC to consider the risks presented to humans and other organisms through the transport of wastes. Although DTSC has indicated that there will be no net effect on the total volumes of waste that are classified as hazardous, special, or nonhazardous (DTSC, personal commun., September 10, 1998), there might be substantial changes in the pattern of movement of wastes. The consequences of such changes should be considered in DTSC's proposed system. If there is substantial additional transport of wastes to more distant landfills (for example, class III landfills are to have a 10-km buffer zone), then there might be a larger risk resulting from the additional transport. The increase in the number of miles traveled might lead to an increase in the number of vehicular accidents, resulting in possible personal injuries and a potential increase in waste spills.

Changes in the pattern of waste movement also raise the question of the acceptability of trade-offs in short-term versus long-term risk. Reclassifying a non-RCRA nonhazardous solid waste to a non-RCRA hazardous waste might ultimately result in a more secure disposal site (e.g., at a landfill with double synthetic liners and leachate monitoring), thus reducing long-term or chronic risk to the environment. Does that reduction, however, warrant the increased potential for more on-site waste handling, longer transport distances from the point of generation to disposal, or greater waste volumes having to be transported and disposed? These are difficult questions to address, but they are reasonable to consider when incorporating a risk-assessment system into the classification of wastes because some action may reduce risk at one point of the waste-management process and simultaneously increase risk at another. Although these issues are policy related, scientific methods may be used to analyze short-term and long-term risk at different points in the waste management system, and can allow multipathway, multimedia comparisons of risk at one waste-management location versus another. The committee supports the use of the multipathway, multimedia risk analysis, but points out that this is currently only being proposed for the location of disposal.

PROGRAM EVALUATION

An essential part of any regulatory program should be a method for determining the success of the program. DTSC's proposed approach fails to incorporate a mechanism to qualitatively or quantitatively monitor the success of the new approach. DTSC provides a process to categorize waste as hazardous or not based on risk estimates associated with specific waste components. The RSU process, which is driving DTSC's review and changes, has as one of its guiding principles the elimination or modification of regulations that are duplicative, ineffective, or do not provide needed protection or information. That goal will be difficult to meet without a feedback mechanism in the program to determine success.

In Chapter 4, the committee criticized DTSC's lack of a mechanism to validate the various attributes of the models that are used to calculate DTSC's regulatory thresholds. On a larger scale, determining the results of the proposed non-RCRA hazardous-waste program by inclusion of a meaningful evaluation system would lend significant credibility to DTSC's desire to meet the objectives of the RSU. To validate the intended effects on protection of human health and other organisms as well as the environment and the regulated community, DTSC should consider the development of a formal, periodic evaluation process.

At a minimum, DTSC should develop a tracking system that provides the data necessary to allow a meaningful assessment of the success of its approach. The development and implementation of a minimal database designed to track various programmatic effects should include the following:

- The actual increase or decrease of waste volume for each non-RCRA hazardous, special, and nonhazardous waste affected by this program.
- Changes in waste movement frequency and routing.
- The use or nonuse of the variance system that DTSC is increasingly relying on to allow program flexibility.

The scientific and regulated communities, as well as the general public, are aware that local, state, and federal regulations are continually modified as new risks are identified or duplication is being eliminated. As a result, the adoption of a regulatory system that has the flexibility to respond to such changes is critical.

References

Burmaster, D.E., and K.M. Thompson. 1995. Backcalculating cleanup targets in probabilistic risk assessments when the acceptability of cancer risk is defined under different risk management policies. Hum. Ecol. Risk Assess. 1(1):101-120.

Burmaster, D.E., K.J. Lloyd, and K.M. Thompson. 1995. The need for new methods to backcalculate soil cleanup targets in interval and probabilistic cancer risk assessments. Hum. Ecol. Risk Assess. 1(1):89-100.

California Department of Human Services. 1981. California Assessment Manual for Hazardous Wastes. Sacramento, CA: California Department of Health Services.

DTSC (Department of Toxic Substances Control). 1997. Introduction to Regulatory Structure Update: April 10, 1997. State of California, Environmental Protection Agency, Hazardous Waste Management Program.

DTSC (Department of Toxic Substances Control). 1998a. Risk-Based Criteria for Non-RCRA Hazardous Waste: A Report to the National Research Council Introducing Proposed Changes to the Definition of Hazardous Waste in the California Code of Regulations. California Environmental Protection Agency, Science, Pollution Prevention and Technology Program. February 27, 1998. 1,494 pages.

DTSC (Department of Toxic Substances Control). 1998b. A Risk-based Regulatory System: Revision of California Non-Resource Conservation And Recovery Act (RCRA) Waste Classification Regulation. Spring 1998.

EPA (U.S. Environmental Protection Agency). 1989. Risk Assessment Guidance for Superfund. Volume 1: Human Health Evaluation Manual (Part A). Office of Emergency and Remedial Response. EPA/540/1-89/002.

EPA (U.S. Environmental Protection Agency). 1990a. Exposure Factors Handbook. Office of Health and Environmental Assessment. EPA/600/8-89/043.

EPA (U.S. Environmental Protection Agency). 1990b. Fugitive Dust Model (FDM) User's Guide (Revised). Volume 1. User's Instructions. Prepared by TRC Environmental Consultants for EPA Region 10. EPA/SW/DK-90/041A.

EPA (U.S. Environmental Protection Agency). 1992. User's Guide for the Industrial Source Complex (ISC2) Dispersion Models, Volume I - User Instructions (EPA-450/4-92-008a), Volume II - Description of Model Algorithms (EPA-450/4-92-008b), Volume III - Guide to Programmers (EPA-450/4-92-008c). Office of Air Quality Planning and Standards, Technical Support Division, Research Triangle Park, North Carolina 27711. March 1992. Addendum A (EPA-450/4-92-008), September 1992.

EPA (U.S. Environmental Protection Agency). 1995a. Air/Superfund National Technical Guidance Study Series: Guideline for Predictive Baseline Emissions Estimation for Superfund Sites. Office of Air Quality Planning and Standards. EPA-451/R-96-001. November.

EPA (U.S. Environmental Protection Agency). 1995b. User's Guide for the Industrial Source Complex (ISC3) Dispersion Models: Volume I - User Instructions (EPA-454/B-95-003a), Volume II - Description of Model Algorithms (EPA-454/B-95-003b). Office of Air Quality Planning and Standards, Emissions, Monitoring, and Analysis Division, Research Triangle Park, North Carolina 27711. September 1995. Addendum: User's Guide for the Industrial Source Complex (ISC3) Dispersion Models, Volume I - User Instructions. December 1998.

EPA (U.S. Environmental Protection Agency). 1996. An SAB Report: Review of a methodology for establishing human health and ecologically based exit criteria for the hazardous waste identification rule (HWIR). EPA-SAB-EC-96-002. May 1996. With a cover letter dated May 22, 1996 from Dr. Genevieve M. Matanoski, Chair Executive Committee, U.S. Environmental Protection Agency, Science Advisory Board, and Dr. Mark A. Harwell and Dr. Ishwar P. Murarka, Co-Chairs HWIR Subcommittee to Honorable Carol M. Browner, Administrator, U.S. Environmental Protection Agency.

EPA (U.S. Environmental Protection Agency). 1997a. Exposure Factors Handbook. Office of Research and Development, Washington, DC . Vol 1: General Factors EPA/600/P-95/002F-a; Vol 2: Food Ingestion Factors EPA/600/P-95/002F-b; Vol. 3: Activity Factors EPA/600/P-95/002F-c.

EPA (U.S. Environmental Protection Agency). 1997b. Compilation of Air Pollutant Emission Factors. AP-42, Fifth Edition, Volume I: Stationary Point and Area Sources. Available from Clearing House for Inventories and Emissions Factors (CHIEF) website at http://www.epa.gov/ttn/chief/.

REFERENCES

Hsieh, D.P.H., T.E. McKone, F.F. Chiao, R.C. Currie. and L. Kleinschmidt. 1994. Intermedia Transfer Factors for Contaminants Found at Hazardous Waste Sites: Trichloroethylene (TCE). Prepared for The Office of Scientific Affairs, The Department of Toxic Substances Control (DTSC) and the California Environmental Protection Agency in Support of the CalTOX Model, December 1994. Draft Final Report.

Ludwig, J.P., J.P. Giesy, C.L. Summer, W.W. Bowerman, R. Aulerich, S. Bursian, H.J. Auman,. P.D. Jones, L.L. Williams, D.E. Tillitt, and M. Gilbertson. 1993. A comparison of water quality criteria for the Great Lakes based on human and wildlife health. J. Great Lakes Res. 19:789-807.

McKone, T.E., and P.B. Ryan. 1989. Human Exposure to Chemicals Through Food Chains: An Uncertainty Analysis. Environ. Sci. Technol. 23:1154-1163.

NIOSH (National Institute of Occupational Safety and Health). 1974. A Recommended Standard: An identification system for occupationally hazardous materials. HEW Publication No. (NIOSH) 75-126. Washington, DC: U.S. Government Printing Office.

NRC (National Research Council). 1994. Science and Judgment in Risk Assessment. Washington, D.C.: National Academy Press.

RTI (Research Triangle Institute). 1995a. Technical support document for the hazardous waste identification rule: risk assessment for human and ecological receptors. Research Triangle Institute for the U.S. Environmental Protection Agency under Contract Nos. 68-D2-0065, 68-W3-0028. August 1995. U.S. EPA Docket F-95-WHWP-FFFFF

RTI (Research Triangle Institute). 1995b. Supplemental technical support document for the hazardous waste identification rule: Risk assessment for human and ecological receptors. Research Triangle Institute for the U.S. Environmental Protection Agency under Contract Nos. 68-D2-0065, 68-W3-0028. November, 1995. U.S. EPA Docket F-95-WHWP-FFFFF

Travis C.C., and A.D. Arms. 1988. Bioconcentration of organics in beef, milk and vegetation. Environ. Sci. Technol. 22:271-274

Appendix A

Biographical Information on the Committee on Risk-Based Criteria for Non-RCRA Hazardous Waste

ROGENE HENDERSON *(Chair)*, Lovelace Respiratory Research Institute, Albuquerque, New Mexico

Rogene Henderson is a senior scientist in the Toxicology Division of the Lovelace Respiratory Research Institute in Albuquerque, New Mexico. She received her B.A. from Texas Christian University and her Ph.D. in chemistry from the University of Texas at Austin. Dr. Henderson is a diplomate of the American Board of Toxicology. She has served on many NRC committees, including the Committee on Epidemiology of Air Pollutants and the Committee on Biologic Markers. Dr. Henderson chaired the NRC's Committee on Toxicology (COT) from 1992-1998. During that period she also chaired COT's Subcommittee on Pulmonary Toxicology, its Subcommittee on PELs for Military Jet Fuels, and its Subcommittee to Review the Army's Toxicologic Risk Assessment of Zinc-Cadmium Sulfide. Dr. Henderson is currently a member of the NRC Board on Environmental Studies and Toxicology.

MARK W. BELL, Parsons Brinckerhoff Energy Services, Inc., Denver, Colorado

Mark W. Bell is assistant vice president and Denver area manager of Parsons Brinckerhoff Energy Services, Inc. He received his M.S. in environmental science from the University of Colorado. He is a member of the Association of Ground Water Scientists and Engineers and the Colorado Ground Water Association. He works as an environmental scientist, and he has covered environmental studies and remediation, regulatory compliance, and program/project management. As area manager of the Denver office, he is responsible for managing the firm's environmental consulting operations in the western U.S., including property transfer, environmental site assessment, environmental impact assessment, cleanup/mitigation development and implementation, environmental permitting, and design/construction support.

JOSEPH F. BORZELLECA, Virginia Commonwealth University, Richmond, Virginia

Joseph F. Borzelleca is emeritus professor of pharmacology and toxicology in the Department of Pharmacology and Toxicology at Virginia Commonwealth University, Medical College of Virginia. He earned his Ph.D. in pharmacology from Thomas Jefferson University Medical College. His research interests include the toxicology of substances of economic importance (e.g. food additives, pesticides, water contaminants) and biodisposition of chemicals including drugs. Dr. Borzelleca has served on several NRC committees including the Committee on Toxicology, the Committee on Food Additives Survey Data, the Committee on Toxicology's Subcommittee on Disinfectants, and the Committee on Safe Drinking Water.

EDWIN H. CLARK, II, Clean Sites, Inc., Washington, DC

Edwin H. Clark is president of Clean Sites Inc., in Washington, DC. He is the former Secretary of Natural Resources and Environmental Control for the State of Delaware, Vice President of the Conservation Foundation, and Associate Assistant Administrator for pesticides and toxic substances in the U.S. Environmental Protection Agency. He holds a Ph.D. in applied economics from Princeton University. He has served as a member of the NRC Board on Environmental Studies and Toxicology.

EDMUND A.C. CROUCH, Cambridge Environmental, Inc., Cambridge, Massachusetts

Edmund A.C. Crouch is a senior scientist for Cambridge Environmental, Inc., and an Associate of the Department of Physics at Harvard

University. Dr. Crouch holds a B.A. in Natural Sciences (Theoretical Physics) and a Ph.D. in High Energy Physics, both from Cambridge University, United Kingdom. Dr. Crouch has published widely in the areas of environmental quality, risk assessment, and presentation and analysis of uncertainties. He has co-authored a major text in risk assessment, *Risk/Benefit Analysis*. Dr. Crouch has served as an advisor to various local and national agencies concerned with public health and the environment. He has written computer programs for the sophisticated analysis of results from carcinogenesis bioassays; has developed algorithms (on the levels of both theory and computer implementation) for the objective quantification of waste site contamination; and has designed Monte Carlo simulations for purposes of fully characterizing uncertainties and variabilities inherent in health risk assessment. Dr. Crouch currently serves as a member of the NRC Committee on the Health Effects of Waste Incineration.

JOHN P. GIESY, Michigan State University, East Lansing, Michigan
John P. Giesy is Distinguished Professor of Zoology in the College of Natural Science at Michigan State University. He also holds appointments in the Institute for Environmental Toxicology and National Food Safety and Toxicology Center. He received his Ph.D. in limnology from Michigan State University in 1974. His research interests include cycling of heavy metals, uptake and availability of heavy metals in aquatic systems, aquatic toxicology, and pesticides. He is the author of over 230 books, book chapters, and journal publications. Currently, he serves on the National Research Council's Committee on Hormonally Active Agents in the Environment and the National Academy of Medicine's panel on the ecological effects of dioxins in Vietnam.

P. BARRY RYAN, Emory University, Atlanta, Georgia
P. Barry Ryan is professor, Exposure Assessment and Environmental Chemistry in the Department of Environmental and Occupational Health at the Rollins School of Public Health of Emory University with a joint appointment as Professor in the Department of Chemistry. Before joining the faculty at Emory in 1995, Dr. Ryan was Associate Professor of Environmental Health at Harvard School of Public Health. He earned his Ph.D. in computational chemistry from Wesleyan University. Research conducted by Dr. Ryan focuses on multimedia, multi-pollutant human exposure assessment and non-traditional pathways of exposure.

He has authored many journal articles and conference papers in his field of expertise.

JAMES N. SEIBER, U.S. Department of Agriculture, Albany, California
James N. Seiber is director of the Western Regional Research Center, U.S. Department of Agriculture. He earned his Ph.D. in organic chemistry from Utah State University. His research interests include environmental analysis and fate of biologically active chemicals, particularly pesticides, industrial byproducts and plant-derived poisons. He was a member of the NRC Committee on Risk Assessment of Hazardous Air Pollutants and Pesticides in the Diets of Infants and Children, and is currently a member of the Committee on Future Role of Pesticides in Agriculture.

CURTIS C. TRAVIS, Project Performance Corporation, Knoxville, Tennessee
Curtis C. Travis is Vice President at Project Performance Corporation. He received his Ph.D. in applied mathematics from the University of California, Davis. As vice president, he is responsible for providing information, engineering and environmental support to government and private sector clients. Previously, he was director of the Center for Risk Management at the Oak Ridge National Laboratory. He has served on several NRC committees including the Subcommittee on Contaminant Plumes, the Committee on Remedial Action Priorities for Hazardous Waste Sites, and the Panel to Review Planned DOE Disposal of Radioactive Waste in Single Shell Tanks at Hanford.

Appendix B

DTSC Issues

DTSC INDICATED in its documentation to the NRC committee that there were specific issues on which DTSC would appreciate committee input (DTSC 1998a, p. 10-12). The committee has provided a discussion of these issues in the main text of the report; this appendix refers the reader to the appropriate chapter and section for each issue. (The Reference Tab and page numbers refer to the DTSC report, 1998a.)

Specific DTSC Issues	Chapter Reference
TTLC ISSUES	
1 Model Selection: Is it appropriate to use a multimedia, multipathway risk assessment model to establish concentration limits on organic chemical constituents in various classes of waste? *Reference Tab 3 Appendix 3 (pg 50) and Tab 4a (pg 76)*	Chapter 2 - Multimedia & Multipathway Risk Assessment
2 Is it appropriate to use a simple exposure model to establish concentration limits on inorganic chemical constituents in various classes of waste? *Reference Tab 3 Appendix 3 (pg 50) and Tab 4c (pg 784)*	Chapter 3 - Analysis of Scenarios and Modeling; Chapter 4 - Preliminary Endangerment Assessment

APPENDIX B: DTSC ISSUES

Specific DTSC Issues	Chapter Reference
3	Is it appropriate to use a lead uptake/blood lead model to establish concentration limits on lead in various classes of waste? *Reference Tab 3, Appendix 3 (pg 50) and Tab 4b (pg 775)*	Chapter 4 - LeadSpread
4	Scenario Selection: Are the waste management worker and the nearby resident scenarios appropriate to represent populations potentially exposed to lower-tier hazardous wastes? The exposures of these populations would be used to establish a "bright line" to separate wastes which are subject to pre-disposal requirements and must be disposed of in RCRA subtitle C (hazardous waste) landfills from those which are subject to reduced pre-disposal and can be safely disposed of in RCRA subtitle D (MSW) facilities? *Reference Tab 3, Appendix 3 (pg 50)* The Department recognizes that some wastes may be disposed of in ways that have less potential exposure. Soil is an example of a waste that has a high probability of being disposed of in a way the results in less dilution and therefore potentially greater exposure.	Chapter 3 - Scenarios: Connection to Policy
5	All assessments of carcinogenic risk are based on estimated exposure to a composite person, having characteristics between those of a child and those of an adult. CalTOX also uses this approach for non-carcinogenic effects, while the PEA-based model and the Lead Risk Assessment Spreadsheet assesses exposure to children and adults separately for non-carcinogenic effects. Which approach is appropriate for carcinogens? For non-carcinogens?	Chapter 4
6	Pathways and Parameters: Do the exposure pathways evaluated for workers and residents reasonably represent the exposure potential for these groups of people? *Reference Tabs 4a (pg 76), 4b (pg 775), and 4c (pg 784)*	Chapter 3 - Scenarios: Connection to Policy

Specific DTSC Issues		Chapter Reference
7	In the original submission it was assumed that 30% of a worker's skin is exposed. The 30% figure used in the assessment falls between worst case and typical case. The latter would involve about 13% exposure. The Department was able to obtain only anecdotal information about landfill workers' attire. Is 30% a reasonable mean value for fraction of skin exposed?	Beyond the scope of the committee's task; however, see Chapter 4 - Parameter Selection within Specific Models for discussion of landfill worker scenario and model parameters
8	Although daily cover of active disposal areas is required and only a limited area of a landfill may be active at any given time, no cover was assumed in modeling the landfill. This may be less important for vapors which are expected to escape through gas collection systems regardless of daily cover or capping. With adequate data, a distribution of uncovered areas for California landfills and waste piles could theoretically be developed. However, DTSC used conservative modeling parameters partly because of data limitations and also because workers and residents may be exposed to vapors and particles from waste piles, which may not have daily cover.	Chapter 3 - Scenarios: Completeness and Coverage
9	The degree of overestimation of exposure from a landfill source may be considerable for particulate pollutants because their release should be significantly reduced by daily cover or a cap. The degree of overestimation of exposure from a landfill source would likely be much less for vapors, because advective transport is likely to result in vapor emissions, usually through a gas collection system, regardless of daily cover or permanent cap. However, only a fraction of the landfill would be producing gases at any given time.	Chapter 3 - Mathematical Models and Their Implementation; Chapter 4 - Preliminary Endangerment Assessment, CalTOX
10	No dust dilution was assumed for PEA and Lead spreadsheet modeling, i.e., all respirable particulates are assumed to come from the waste. Monitoring data for California land-	Chapter 4 - Preliminary Endangerment Assessment

APPENDIX B: DTSC ISSUES 133

Specific DTSC Issues	Chapter Reference
11	fills and waste piles were sought, but most existing data are based on complaint follow-ups, and would be highly biased. U.S. EPA has withdrawn the oral cancer potency factor for beryllium. Treating beryllium as a carcinogen only by the inhalation route would result in about a six-fold increase in the beryllium TTLC. Should beryllium be treated as an oral carcinogen?	(Not discussed)
12	Other Issues: Is the proposed stepwise approach to ensuring that TTLCs based on human toxicity will protect non-human biota reasonable? *Reference Tab 3, Appendix 3 (pg 50) and Tab 4d (pg 844)*	Chapter 3 - Ecological Scenario
13	Is the proposed use of two times the estimated quantitation level (EQL) in lieu of a TTLC when the calculated concentration is less than the EQL reasonable? *Reference Tab 5b (pg 864)*	Chapter 4 - Analytical Methods
14	Is the proposed approach to considering ambient levels of elements and compounds in setting TTLCs appropriate? *Reference Tab 5a (pg 859)*	Chapter 4 - Analytical Methods
SERT ISSUES:	
15	SERTs are based on the lowest of (1) health-based concentrations calculated by DTSC, (2) California Maximum Contaminant Levels established by the Department of Health, or (3) U.S. EPA Ambient Water Criteria for the protection of aquatic life. Is this paradigm appropriate? *Reference Tab 3, Appendix 2 (pg 43)*	Chapter 4 - Soluble or Extractable Regulatory Thresholds
16	Is the dilution/attenuation factor of 100 for the SERTs appropriate? *Reference Tab 3, Appendix 2 (pg 43)*	Chapter 4 - Soluble or Extractable Regulatory Thresholds
17	Is the use of a factor to account for the retardation of leakage from the landfill due to the	Chapter 4 - Soluble or Extractable

Specific DTSC Issues	Chapter Reference
	presence of a synthetic liner appropriate? *Reference Tab 3 Appendix 2 (pg 43)* Was the liner protection factor calculated appropriately? *Reference Tab (12) (pg 1485)*	Regulatory Thresholds
18	The current STLCs and proposed SERTs are based on an assessment of the environmental effects of leachates from solid wastes in a landfill, but apply also to liquid wastes that are not disposed of in a landfill. Should the effects of released liquids be assessed separately? *Reference Tab 2, Appendices 2 (pg 43) and Tab 6 (pg 869)*	Chapter 2 - Implementation Practicality and Evaluation
19	Is laboratory extraction of toxic constituents from solid wastes with municipal solid waste leachates a reasonable standard against which to compare the performance of laboratory extraction protocols such as the WET, the TCLP, and the SPLP? Is the Department's conclusion that the WET is not a better estimator of leaching potential than the TCLP justified? *Reference Tab 6 (pg 869), Tab (9) (pg 1046), Tab (10) (pg 0178) and Tab (11) (pg 1255)*	Chapter 4 - Analytical Issues
20	We concluded that the extraction by municipal solid waste leachates of elements capable of forming oxyanions was not simulated by the WET or by the TCLP. Is this a reasonable conclusion? *Reference Tab 3 Appendix 2 (pg 43) and Tab 6 (pg 869), Tab (9) (pg 1046), Tab (10) (pg 0178) and Tab (11) (pg 1255)*	Chapter 4 - Analytical Issues
21	In the current proposal, ingestion of ground water is the only human exposure pathway evaluated. Does this approach seriously underestimate risk? *Reference Tab 3 Appendix 2 (pg 43)*	Chapter 3 - Scenarios: Connection to Policy
22	Is the use of the 90th percentile concentration of arsenic in monitored Calif. drinking-water supplies appropriate as a basis for the arsenic SERT? *Reference Tab 3 Appendix 2 (pg 43)*	Chapter 4 - Soluble or Extractable Regulatory Thresholds

Specific DTSC Issues	Chapter Reference
23	Is the proposed use of two times the estimated Quantitation level (EQL) in lieu of a SERT when the calculated concentration in less than the EQL reasonable? *Reference Tab 5b (pg 864)*	Chapter 4 - Analytical Issues
24	Can the STLC values for arsenic, antimony, molybdenum, selenium, and vanadium be adjusted in a scientifically sound manner to compensate for the extent of inaccuracy of the testing procedure (TCLP) for those constituents?	Chapter 4 - Analytical Issues
ACUTE TOXICITY ISSUES	
25	Is the Department's proposed approach to setting acute oral, dermal, and inhalation toxicity thresholds reasonable? *Reference Tab 3 Appendix 4 (pg 72)*	Chapter 4 - Toxicity Tests

Appendix C

List of Public Access Materials Received by the NRC Committee on Risk-Based Criteria for Non-RCRA Hazardous Waste

1. Letter to Jesse Huff, Department of Toxic Substances Control from Jeff Sickenger, Western States Petroleum Association, Brian E. White, California Chamber of Commerce, Stephen P. Piatek, Western Independent Refiners Association, Jot Condie California Manufacturers Association, and Robert W. Lucas, California Council for Environment and Economic Balance. Re: Waste Classification (April 30, 1998, 22 pp).
2. Letter to Jan Radimsky, California Department of Toxic Substances Control from Thomas J. LeBrun, County Sanitation Districts of Los Angeles County. Re: Proposed Hazardous Waste Classification (Sept. 4, 1997, 2 pp).
3. Four Letters to Jesse Huff, Department of Toxic Substances Control from Robert W. Lucas, California Council for Environmental and Economic Balance. Re: Hazardous Waste Characterization (Nov. 17, 1997, 10 pp) (Jan. 8, 1998, 5 pp); Arsenic (Feb. 27, 1998, 6 pp); SERTs (April 24, 1998, 7 pp).
4. Letters to Jesse Huff, Department of Toxic Substances Control from Joseph W. Massey, Institute of Scrap Recycling Industries, Inc. Re: Waste Classification/Management (Nov. 25, 1997, 5 pp); Letter to James Carlisle, Department of Toxic Substances Control from

Joseph W. Massey, Institute of Scrap Recycling Industries, Inc. Re: Regulatory Structure Update, Draft Concept paper RSU Task D.3, California's Non-RCRA Waste Classification System, Analysis and Proposed Revisions Dated Feb. 12, 1997 (Mar. 26, 1997, 4 pp); Letter to California Environmental Protection Agency from Jeffrey P. Neu, Hugo Neu-Proler Company. Re: Regulatory Improvement Initiative with attachments (Dec. 6, 1995, 16 pp).

5. Letter to Jesse Huff, Department of Toxic Substances Control from David McKinley, Industrial Environmental Association. Re: DTSC RSU Waste Classification Proposal (Nov. 21, 1997, 2 pp).

6. Four letters to Jesse Huff, Department of Toxic Substances Control from June Christman, Western Independent Refiners Association. Re: Response to Your July 28, 1998 Letter, RSU Task B.I.5 Discharges to Publicly Owned Treatment Works (Aug. 13, 1998, 6 pp); RSU - Limited Domestic Sewage Exclusion (May 18, 1998, 3 pp); Regulatory Structure Update (RSU) Waste Characterization (Mar. 10, 1998, 3 pp); RSU - Hazardous Waste Characterization Proposal and Limited Domestic Sewage Exclusion (Jan. 28, 1998, 4 pp).

7. Letter to Jesse Huff, Department of Toxic Substances Control from V. Z. Froman, Department of the Navy, Commander Naval Base. (June 9, 1998, 1 p).

8. Letter to Jesse Huff, Department of Toxic Substances Control from Jeff Sickenger, Western State Petroleum Association. Re: Waste Classification (Nov. 19, 1997, 3 pp, with 5 attachments).

9. Risk-Based Criteria for Non-RCRA Hazardous Waste: A Report to the National Research Council Introducing Proposed Changes to the Definition of Hazardous Waste in the California Code of Regulations [prepared by Human and Ecological Risk Division and Hazardous Materials Laboratory of the Science, Pollution Prevention, and Technology Program, Department of Toxic Substances Control, Environmental Protection Agency, State of California. Feb. 27, 1998, 1494 pp]. Cal/EPA DTSC also provided the following materials to the committee:

(a) The DTSC Waste Classification Proposal Summary & Highlights [Robert Stephens, Deputy Director, California Environmental Protection Agency, Department of Toxic Substances Control, Sept. 10, 1998, 34 pp].

(b) Assembly Bill No. 2784 [Robert Stephens, Deputy Director, California Environmental Protection Agency, Department of Toxic Substances Control, 13 pp].

(c) Final Draft Reports. Intermedia Transfer Factors for Contaminants Found at Hazardous Waste Sites: Vinyl Chloride (VC) [40 pp], Trichloroethylene (TCE) [45 pp], and 2,3,7,8-Tetrachlorodibenzo-p-dioxin (TCDD) [44 pp].

(d) Additional and Replacement Information: CalTOX Technical Reports, modifications to create version 2.3, chemical parameters for TCE, TCDD, and VC (1-p memorandum and disk) [Ned Butler, DTSC Staff Toxicologist, Cal/EPA].

(e) A Risk-Based Regulatory System: Revision of California Non-Resource Conservation and Recovery Act (RCRA) Waste Classification Regulation. DTSC, obtained from web site http://www/calepa.ca.gov/dtsc/rsu/hwtr0623.pdf [58 pp].

(f) Introduction to Regulatory Structure Update and List of Tasks, includes Draft Concept Papers for Tasks D.1, D.2, D.3, and D.4, State of California, Environmental Protection Agency, Department of Toxic Substances Control, April 10, 1997; Unified Program: Questions and Answers, Unified Hazardous Waste and Hazardous Management Regulatory Program, Mar. 28, 1997 [48 pp].

10. Risk Assessment Strategy for the Hazardous Waste Identification Rule (HWIR) [Barnes Johnson, Office of Solid Waste and Emergency Response, U.S. Environmental Protection Agency, Washington, DC, 7 pp].

11. The following materials were received from the Toxics Assessment Group in support of Jane Williams of the California Communities Against Toxics:

(a) Letter to Members, Committee on Risk-Based Criteria for Non-RCRA Hazardous Waste from Jody Sparks, Toxics Assessment Group; Jane Williams, California Communities Against Toxics, Gary Patton, Planning and Conservation League, Caryn Woodhouse, Taylor Bay Associates, and Bonnie Holmes-Gen, Sierra Club, dated Sept. 10, 1998, 9 pp].

(b) Comments on the Draft Concept Paper (Task D4): Preliminary Proposal to Require the Federal Toxicity Leaching Procedure (TCLP) in lieu of the Waste Extraction Test (WET) [letter from Caryn Woodhouse, Taylor Bay Associates, to Dr. James Carlisle, DTSC, dated Jan. 2, 1996, 10 pp].

(c) Comments on the Draft Concept Paper (Task D1 and D3): Evaluation of RCRA Listed Wastes and the RCRA Toxicity Characteristic [letter from Jody Sparks, Toxics Assessment

Group and Caryn Woodhouse, Taylor Bay Associates, to Dr. James Carlisle, DTSC, dated Feb. 8, 1996, 10 pp].

(d) Comments on the Regulatory Structure Update Program [letter from Jody Sparks, Toxics Assessment Group, to Jesse Huff, DTSC, dated Mar. 13, 1996, 4 pp].

(e) Adequacy of DTSC Response to RSU: Comments in Jan. 16, 1996 Memorandum [letter from Jody Sparks, Toxics Assessment Group and Caryn Woodhouse, Taylor Bay Associates, to Dr. James Carlisle, DTSC, dated April 1, 1996, 19 pp].

(f) Comments on the Draft Concept Paper, RSU Task #C.3: Soils: natural Contamination, Petroleum Contamination, Replacement [letter from Jody Sparks, Toxics Assessment Group and Caryn Woodhouse, Taylor Bay Associates to Jan Radimsky, DTSC, dated May 2, 1996, 6 pp].

(g) Comments on the Regulatory Structure Update program: Laboratory Comparison Study [letter from Bonnie Holmes-Gen, Sierra Club, Jody Sparks, Toxics Assessment Group, Gary Patton, Planning and Conservation League, and Jane Williams, California Communities Against Toxics, to Robert Stephens, DTSC, dated June 20, 1996, 4 pp].

(h) Comments on the Draft Concept Paper, RSU Task #D.3: Evaluation of California's non-RCRA Hazardous Waste System, Part A: STLCs and Protection of Water Resources, Aug. 17, 1996 [letter from Jody Sparks, TAG and Caryn Woodhouse, Taylor Bay Associates, to Jan Radimsky and Jim Carlisle, DTSC, dated Sept. 9, 1996, 6 pp].

(i) Comments on the Draft Concept Paper, RSU Task #D.4: Extraction Test and the RSU Extraction Test Report by Hazardous Materials Laboratory [letter from Jody Sparks, TAG and Caryn Woodhouse, Taylor Bay Associates, to Jan Radimsky and Jim Carlisle, DTSC, dated Sept. 9, 1996, 15 pp].

(j) AB 1220 Regulations - Impact of Deregulated Hazardous Wastes [letter from Jody Sparks, Toxics Assessment Group and Caryn Woodhouse, Taylor Bay Associates, to Members of the Integrated Waste Management Board and Members of the of the State Water Resources Control Board, dated Sept. 11, 1996, 3 pp].

(k) Phase II RSU Extraction Test Project [letter from Jody Sparks, Toxics Assessment Group, Jane Williams, California Communities Against Toxics, Stephen Johnson, HEWM, and Caryn

Woodhouse, Taylor Bay Associates, to Robert Stephens, DTSC, dated Oct. 25, 1996, 8 pp].
(l) Phase II RSU Extraction Test Project [letter from Jody Sparks, Toxics Assessment Group, to Robert Stephens, DTSC, dated Dec. 12, 1996, 2 pp].
(m) Comments on the Draft Concept Paper, RSU Task #D.3: California's nonRCRA Hazardous Waste System - Analysis and Proposed Revisions [letter from Jody Sparks, Toxics Assessment Group and Caryn Woodhouse, Taylor Bay Associates, to Jan Radimsky and Jim Carlisle, DTSC, dated April 3, 1997, 17 pp].

12. Waste Classification [Michael Lakin, ICF Kaiser, representing the Western States petroleum Association, California Chamber of Commerce, Council for Economic and Environmental Balance, California Manufacturers Association, Western Independent Refiners Association, 25 pp].
   (a) Proposed Presentation on Waste Classification (unabridged) [Michael Lakin, ICF Kaiser, 50 pp].

13. (a) Discharges to POTWs [Linda M. Shadler, Supervising Civil Engineer, County Sanitation Districts of Los Angeles County, 4 pp].
   (b) Selenium Chemistry in POTWs [Brent C. Perry, Project Engineer, County Sanitation Districts of Los Angeles County, Aug. 3, 1998, 4 pp].

14. Reference Materials for National Academy of Sciences Public Meeting, NRC Committee on Risk-Based Criteria for Non-RCRA Hazardous Waste [Charles A. White, Director of Regulatory Affairs, Waste Management, Sacramento, CA, 5 pp].

15. Review of Cal/EPA's proposed "Risk-Based Criteria for Non-RCRA Hazardous Wastes" [Paul W. Abernathy, Consultant to Mercury Technologies International/Advanced Environmental Recycling Corporation, Hayward, CA, 7 pp].

16. Comments submitted by Victor Hanna, representing the City of Los Angeles, Bureau of Sanitation, to NAS committee on Sept. 10, 1998, 1 p.

17. Waste Classification - Focus on Real Risk [letter from Brian E. White, Air & Waste Management, California Chamber of Commerce, to Jesse Huff, Director, DTSC, dated Mar. 9, 1998, 4 pp].

18. Comments on Draft Concept Paper for RSU Task D-3 [letter from Michelle Corash, Morrison & Foerster, dated May 1, 1997; cover letter from Brooke Ashworth, Environmental Analyst, Morrison &

Foerster, to Roberta Wedge, NRC, dated Sept. 2, 1998, 7 pp].
19. Waste Deregulation [e-mail from David D. Miller, Vice President, California Association of Professional Scientists to Roberta Wedge, NRC, dated Sept. 9, 1998, 1 p].
20. Reclassifying Hazardous Waste [e-mail from Christopher J. Voight, Staff Consultant, California Association of Professional Scientists to Roberta Wedge, NRC, dated Sept. 9, 1998, 1 p].
21. Risk Assessment - A Science Based Approach. A Symposium (transcript of symposium, roster, and program). Held by Southern California Coalition for Pollution Prevention and the California Environmental Protection Agency, Department of Toxic Substances Control, May 16, 1997 [cover letter from Christopher Campbell, Executive Director, Southern California Coalition for Pollution Prevention to Roberta Wedge, NRC, Sept. 15, 1998, 202 pp].
22. Bolger, P.M., Carrington, C.D., Capar, S.G., and Adams, M.A. Undated. Reductions in Dietary Lead Exposure in the United States. U.S. Food and Drug Administration, Division of Toxicological Review and Evaluation, Division of Contaminants Chemistry and Division of Food Chemistry and Technology. 26 pp.
23. Provisional Tolerable Exposure Levels for Lead [memo from Contaminants Team, Division of Toxicological Review and Evaluation, Department of health and Human Services, to Elizabeth Campbell, Division of Regulatory Guidance, dated Nov. 16, 1990, 10 pp].
24. Chaney, R.L., Mielke, H.W., and Sterrett, S.B. 1988. Speciation, Mobility, and Bioavailability of Soil Lead. Submitted to Environ. Geochem. Health. 23 pp.
25. California Non-RCRA Hazardous Waste Classification Proposal [comments from Jon B. Marshack, California Regional Water Quality Control Board, Central Valley Region, to Roberta Wedge, NRC, dated Oct. 7, 1998, 8 pp].
26. Chaney, R.L., Sterrett, S.B., and Mielke, H.W. 1984. The potential for heavy metal exposure from urban gardens and soils. Pp. 37-84 in Proceedings of the Symposium on Heavy Metals in Urban Gardens, J.R. Preer, ed. Agricultural Experiment Station, University of the District of Columbia, Washington.
27. Comments for NRC Committee on Risk-Based Criteria for Non-RCRA Hazardous Waste [letter from Gordon E. Hart, of Paul, Hastings, Janofsky & Walker LLP, to Roberta Wedge, NRC, dated Oct. 9, 1998, 5 pp].
28. Department of Toxic Substances Control Responses to 86 Questions

from the NRC Committee on Risk-Based Criteria for Non-RCRA Hazardous Waste, dated Oct. 9, 1998. 45 pp plus 7 attachments.
29. Risk-Based Criteria for Non-RCRA Hazardous Waste, BEST-K-98-03-A, 10 Micron Exclusion [letter from Paul Neil, RTP Environmental Associates, Inc., to Roberta Wedge, NRC, dated Oct. 2, 1998, 2 pp].
30. Comments on Department of Toxic Substances Control: California Non-RCRA Hazardous Waste Classification Proposal [letter from Denis L. Brown, Texaco North American Production, to Roberta Wedge, NRC, dated Nov. 18, 1998, 2 pp].
31. Comments on Regulatory Structure Update [letter from Mark Posson, Lockheed Martin Corporation, to Peter M. Rooney, secretary for environmental protection, Cal/EPA, dated Oct. 9, 1998, 3 pp].
32. Comments on DTSC Hazardous Waste Classification System [from Jeff Sickenger, Western States Petroleum Association, Nov. 13, 1998, 70 pp].
33. Department of Toxic Substances Control Supplemental Questions and Answers (pertaining to extraction test methodology) [from DTSC Cal/EPA, dated Nov. 19, 1998, 2 pp].
34. There Are Two Basic Requirements for Hazardous Waste (HW): Management and Disposal [overheads for presentation by Michael Lakin, ICF Kaiser, Nov. 20, 1998, 12 pp].
35. Scenarios, Benefits and Costs [overheads for presentation by Kirk T. Larson, representing San Diego Industrial Environmental Association, Nov. 20, 1998, 11 pp].
36. SCE Comments to NAS Public Meeting on RSU Waste Characterization Project [comments from David Kay, Southern California Edison Co., Nov. 20, 1998, 2 pp].
37. Proposed California Waste Classification System [letter from Charles A. White, Waste Management, to Peter Rooney, secretary for environmental protection, Cal/EPA, dated Oct. 7, 1998, 37 pp].
38. Department of Toxic Substances Control Review Questions and Answers [letter and attachment from Robert Stephens, DTSC, Cal/EPA, to Rogene Henderson, chair, NRC committee, undated, 3 pp].
39. Silicon Valley Manufacturing Group comments to NRC committee [letter from Chris Elias, Silicon Valley Manufacturing Group to Roberta Wedge, NRC, dated Dec. 10, 1998, 8 pp].
40. Non-RCRA Hazardous Waste Criteria for Contaminated Soils and Soil Amendments [letter and attachment from Robert Stephens,

DTSC, Cal/EPA, to Rogene Henderson, chair, NRC committee, and Roberta Wedge, NRC staff, dated Dec. 30, 1998, 2 pp, and letter response from Rogene Henderson to Robert Stephens, dated Jan. 22, 1999, 1 p].

41. Waste Classification - NAS Review [comments from A.S. Chater, Hugo Neu-Proler Company to Rogene Henderson, chair, NRC committee, e-mail dated Feb. 2, 1999, 2 pp].

42. Minor issues of clarification [questions from Roberta Wedge, NRC staff to Ned Butler, DTSC, and response from DTSC, dated Jan. 26, 1999, 2 pp].

Appendix D

California Environmental Protection Agency
Department of Toxic Substances Control (DTSC)

Letter of Introduction, Overview, Concept Paper,
and Appendices 1-4 from DTSC Report

Risk-Based Criteria for Non-RCRA Hazardous Waste

Volume 1 of 2

A Report to the National Research Council Introducing Proposed Changes to the Definition of Hazardous Waste in the California Code of Regulations

Prepared by

Human and Ecological Risk Division and
Hazardous Materials Laboratory
of the
Science, Pollution Prevention, and Technology Program
Department of Toxic Substances Control
Environmental Protection Agency
State of California

February 27, 1998

Risk-Based Criteria for Non-RCRA Hazardous Waste

Table of Contents (revised, August 26, 1998)

Tab*	Title	Page
	Volume I	
1	Overview and Letter of Introduction	1
2	Issues	10
3	Concept Paper: California's Non-RCRA Waste Classification System: Analysis and Proposed Revisions (with 4 Appendices)	13
4	TTLC Risk Assessment Models	75
4a	Documents Describing Results and Input Values	76
4a1	CalTOX Adaptations for Derivation of Exit and Upper TTLC Criteria	78
4a2	CalTOX Version 2.3: Description of Modifications and Revisions	105
4a3	CalTOX: A Multimedia Total Exposure Model for Hazardous-Waste Sites	181
4a4	Analysis of Results and Input Values of the Upper and Lower TTLCs Computed Using the CalTOX Model	520
4a5	The Distribution of California Landscape Variables for CalTOX	555
4a6	Parameter Values and Ranges for CALTOX	591
4a7	Draft Final Report: Intermedia Transfer Factors for Contaminants Found at Hazardous Waste Sites	647
	I. Vinyl chloride	647
	II. 2,3,7,8-Tetrachlorodibenzo-p-dioxin	687
	III. Trichloroethylene	730
	Volume 2	
4b	The DTSC Lead Risk Assessment Spreadsheet	775
4c	The Preliminary Endangerment Assessment Model	784
4d	De Novo Ecological Risk Assessments	844
5	Detection Limits and Background Level	858
5a	Ambient and Background Concentrations	859
5b	Analytical Issues	864
6	Concept Paper: Evaluation of the Suitability of the Federal Toxicity Characteristic Leaching Procedure (TCLP) in Lieu of California's Waste Extraction Test (WET)	869
(7)	Comparison of California's Waste Extraction Test (WET) and the U.S. EPA's Toxicity Characteristic Leaching Procedure (TCLP)	880
(8)	Leaching Potential of Persistent and Bioaccumulative Toxic Substances in Municipal Solid Waste Landfills	964
(9)	Comparison of Short-term Extraction Tests with Extraction Using Municipal Solid Waste Leachates	1046
(10)	Supplement to the RSU Extraction Test Project Summary Report: Phase 1	1078
(11)	Supplement to the RSU Extraction Test Project Summary Report: Phase 2	1255
(11a)	Tables for RSU Phase 2	1285
(11b)	Figures for RSU Phase 2	1331
(11c)	Appendices: Commercial Uses of Elements	1397
7 (12)	Subtitle D Landfill Composite Liner Protection Factor	1485

* numbers in () correspond to Cal/EPA's original numbering system and may be referred to elsewhere in the document

APPENDIX D: DTSC REPORT

Cal/EPA

February 27, 1998

Department of
Toxic Substances
Control

Pete Wilson
Governor

400 P Street,
4th Floor
P.O. Box 806
Sacramento, CA
95812-0806

Mr. Raymond Wassel, Program Director
Board on Environmental Studies and Toxicology
Commission on Life Sciences
National Research Council
2101 Constitution Avenue
Washington, D.C. 20418

James M. Strock
Secretary for
Environmental
Protection

Dear Mr. Wassel:

The Department of Toxic Substances Control (DTSC) has undertaken a complete review of our hazardous waste regulatory program. A basic premise of this review is that each regulatory requirement that exceeds the requirements of the federal Resource Conservation and Recovery Act (RCRA) program must be reviewed. Any requirement that is not scientifically justified and/or does not add value must be revised or repealed. This request for peer review is concerned with the waste classification portion of the program and more specifically with the toxicity characteristic.

Existing Waste Classification System

Currently, waste that is not classified as hazardous under the RCRA program is classified as a non-RCRA hazardous waste in California if it exhibits any of four characteristics: ignitability, reactivity, corrosivity, and toxicity. The first two are defined in a manner identical to the corresponding characteristics under RCRA, so only wastes that are RCRA exempt or excluded could be non-RCRA hazardous under these characteristics. The corrosivity characteristic as defined in California regulations is slightly broader than under RCRA by including corrosive solids. The subject of this request for scientific review is the toxicity characteristic. Currently, a California waste is a non-RCRA hazardous waste due to the characteristic of toxicity if it exceeds any of eight criteria:

- It contains total concentrations of any of thirty-seven toxic constituents exceeding total threshold limit concentrations (TTLCs).
- It contains soluble or extractable concentrations or total concentrations in liquid waste of any of thirty-five toxic constituents exceeding soluble threshold limit concentrations (STLCs).
- Its acute oral LD_{50} is less than 2500 mg waste/kg body weight.
- Its acute dermal LD_{50} is less than 4300 mg/kg body weight.
- Its acute inhalation LC_{50} is less than 10,000 ppm.

Mr. Raymond Wassel, Program Director
February 27, 1998
Page 2

- Its aquatic LC_{50} is less than 500 mg/l.
- It contains any of sixteen listed carcinogens at concentrations exceeding 10 mg/kg.
- It has been shown by experience or testing to be hazardous to public health or the environment (a.k.a. the "new threats" criterion).

The STLCs are based on drinking water MCLs or on fish toxicity. Most of the TTLCs are 100 times the STLCs, but a few have been modified because they were thought to be too high or low.

Proposed Changes in Waste Classification

DTSC has proposed new waste classification criteria based on two premises:

- Criteria should be risk-based and should be set at a constant incremental risk threshold, and
- risk thresholds should be graded to allow better tailoring of management requirements to the severity of the hazard.

In keeping with those premises, DTSC has proposed two tiers of TTLCs and Soluble or Extractable Regulatory Thresholds (SERTs) to divide hazardous from special wastes and special wastes from non-hazardous wastes. Likewise, two levels acute oral, dermal, and inhalation toxicity thresholds are proposed. Aquatic toxicity and named carcinogens would remain single-tiered. Vinyl chloride would be removed from the named carcinogens list and added to the SERT and TTLC lists. The "new threats" category would remain unchanged.

The proposed lower-tier SERTs are the lowest of (1) levels calculated by DTSC using distributions of drinking water consumption normalized to body weight, with a dilution/attenuation factor of 100 or (2) 100 times the primary MCL, or (3) 100 times the federal Ambient Water Quality Criterion. Upper-tier SERTs are the product of the lower SERT and a liner protection factor to account for the effect of a composite liner (disposal in a composite-lined landfill meeting RCRA subtitle D standards will be a requirement for special wastes that are to be land-disposed).

The first step in calculating proposed exit-level TTLCs is to determine the lower of (1) human-health-based or (2) ecosystem-based concentrations, both of which were calculated assuming a land application scenario. The human-health-based concentrations for organic chemicals were calculated using CalTOX, a multimedia, multipathway

APPENDIX D: DTSC REPORT   151

Mr. Raymond Wassel, Program Director
February 27, 1998
Page 3

fugacity-based model. For inorganic constituents, a simple three-pathway exposure model (a modified version of the DTSC Preliminary Endangerment Assessment or PEA model) was used. For mercury and selenium, ecosystem-based exit concentrations were calculated by DTSC in conjunction with Cal EPA's Office of Environmental Health Hazard Assessment.

Proposed upper-tier TTLCs assume that wastes not exceeding this tier of thresholds would be disposed of in a lined municipal landfill meeting RCRA Subtitle D standards. Human populations of concern included workers at the landfill or other waste management facility and residents living near a landfill or other waste handling facility. The workers were assessed using the PEA model. Exposure of nearby residents to organic landfill chemicals was assessed using a modified version of CalTOX that incorporates off-site transport. Exposure of nearby residents to inorganic chemicals in wastes was assessed using the PEA model. Both upper and lower candidate TTLCs were also screened, and if necessary, modified to ensure that they are not below analytical quantitation limits or ambient environmental concentrations

SERTs and TTLCs were calculated as distributions rather than point estimates, and the tenth percentile of the distributions was selected as the basis for the proposed regulatory thresholds (tenth percentile estimates of SERTs or TTLCs correspond to the ninetieth percentile estimates of risk). The reference carcinogenic risk level was 10^{-5} for both SERTs and TTLCs. Health benchmarks were California Environmental Protection Agency cancer potency factors and U.S. EPA reference doses for non-carcinogenic effects.

An attempt was made to develop acute toxicity thresholds based on plausible exposures. Proposed acute oral toxicity thresholds were set at ninetieth percentile estimates of child (lower) and adult (upper) inadvertent ingestion of soil normalized to body weight, using EPA Exposure Factors Handbook data. Proposed lower and upper acute dermal toxicity thresholds were set at ninetieth percentile (upper-end) estimates of child and adult dermal exposure to soil, normalized to body weight using EPA Exposure Factors Handbook data. Proposed acute inhalation toxicity thresholds were set in relation to the volatility or respirability of the waste using plausible exposure scenarios. The existing aquatic toxicity threshold appeared to represent a plausible exposure scenario for the lower threshold. The proposed upper aquatic toxicity threshold is the current level divided by a tenth percentile estimate of the liner protection factor.

Mr. Raymond Wassel, Program Director
February 27, 1998
Page 4

Review Package

DTSC is requesting the panel's review of the scientific elements of this waste classification revision package. However, to help the panel focus on those elements the Department considers most critical, we have enclosed a list of issues in the form of questions, which is behind Tab 2. This package contains the following elements, arranged by tab number:

1. An overview of the current California Waste Classification System and proposed changes.
2. Issues - The specific questions for which DTSC is asking the panel's input.
3. Waste classification concept paper.
4. TTLC Risk Assessment Supporting Documents.
5. TTLC Science Policy Supporting Documents.
6. Estimation of soluble or extractable concentrations.
7. Subtitle D Landfill Composite Liner Protection Factor.

Sincerely,

Jesse Huff
Director

Robert D. Stephens, Ph.D
Deputy Director

APPENDIX D: DTSC REPORT   153

1. Overview

California's system for classifying wastes consists of two principal elements. One element is the federal system created under the authority of the Resource Conservation and Recovery Act (RCRA). As an authorized state, California has promulgated regulations mirroring those developed by the U.S. EPA under RCRA. These regulations are not part of this review. The second element is the non-RCRA system, which goes beyond and complements the RCRA system. The Department of Toxic Substances Control (Department) has undertaken an update of its entire non-RCRA waste scheme. The purpose of this update is to re-assess the need for the non-RCRA system, to update the scientific underpinnings of the system as needed, and to simplify the system where possible by eliminating redundancy and overlap, rewording confusing sections of regulation, and eliminating elements that have outlived their usefulness. It is the scientific basis for the proposed revised waste classification system that we are asking you to review.

California's waste classification differs from the federal system in several ways: California has a list of approximately 800 chemicals which, when present in a waste, make it presumptively hazardous. However, a generator can use testing or knowledge to show that a waste containing one or more of the chemicals on the presumptive list does not exhibit a hazardous characteristic. Thus, California's non-RCRA system does not have "listed" hazardous wastes in the same way that the RCRA system does. California's non-RCRA system, like the RCRA system, includes the four hazardous characteristics of reactivity, ignitability, corrosivity, and toxicity. The first two are identical in the RCRA and non-RCRA systems. California's corrosivity characteristic is slightly broader in that it includes corrosive solids. However, the characteristic that is most different between the two systems, and the one that is the subject of this review, is the toxicity characteristic.

California's existing classification system comprises eight criteria for the toxicity characteristic (see table below). The Department proposes to retain all eight of the criteria, but to revise five of them. At present a single tier of thresholds creates a single tier of regulated waste: hazardous waste. A key aspect of the proposed changes is to add a second tier of regulatory thresholds for each of the first five criteria. This would create two tiers of regulated waste. The upper tier, to be called hazardous waste, would include the most toxic wastes (along with corrosive, ignitable, and reactive wastes). The lower tier would include wastes that are moderately toxic. The principal concern for these wastes is that they not be disposed of indiscriminately.

The following table summarizes regulation number, description of the subject matter, decision and location of basis for the decision. The footnotes are numbered from those subsections undergoing the least revision to those undergoing the greatest revision.

CCR* subsection	Subsection description	Decision	Decision basis
66261.24(a)(1)	Federal Definition of Hazardous Waste[1]	retain	See below
66261.24(a)(2)	Total Threshold Limit Concentration (TTLC)[8]	revise	Tab 3: page 7
66261.24(a)(2)	Soluble Threshold Limit Concentration (STLC)[9]	revise	Tab 3: page 9
66261.24(a)(3)	Acute Toxicity- Oral[5]	revise	Tab 3: page 10
66261.24(a)(4)	Acute Toxicity- Oral[6]	revise	Tab 3: page 11
66261.24(a)(5)	Acute Toxicity- Oral[7]	revise	Tab 3: page 12
66261.24(a)(6)	Fish Bioassay[4]	revise	Tab 3: page 13
66261.24(a)(7)	Carcinogens[3]	retain	Tab 3: page 14
66261.24(a)(8)	Experience or Testing ("New Threats")[2]	retain	Tab 3: page 16

*Citation in Title 22 of the California Code of Regulation (CCR)

1. State RCRA programs must conform to federal law.
2. The Experience or Testing (new threats) subsection of the regulation will be retained as is.
3. Because of its economic importance, TTLCs and SERTs were derived for vinyl chloride. Therefore, vinyl chloride will be removed from the carcinogen list and added to the other two lists. The balance of the listed carcinogens will be retained, and if a waste contains any of these 15 chemicals at concentrations exceeding 10 ppm, they will be classified as hazardous.
4. Fish bioassay will be retained, but the single LC_{50} threshold for defining hazardous waste will be replaced with an upper and an exit LC_{50}. The basis of this selection is in Appendix 4 of Tab 3.
5. The single oral LD_{50} value used to determine hazardous from non-hazardous waste will be replaced with an upper and exit LD_{50}. The basis of this selection is described in Appendix 4 of Tab 3.
6. The single dermal LD50 value used to determine hazardous from non-hazardous waste will be replaced with an upper and exit LD_{50}. The basis of this selection is described in Appendix 4 of Tab 3.
7. The algorithm used to determine hazardous from non-hazardous waste based on inhalation toxicity will replaced with two new algorithms. The basis of these algorithms is described in Appendix 4 of Tab 3.
8. See below for a discussion of TTLCs
9. See below for a discussion of SERTs (formerly STLCs)

The changes described in footnotes 1-7 are relatively simple requiring short explanations. The bulk of documentation (Tabs 3-7) pertain to the new approaches proposed for replacing the values for each TTLC and SERT. Each of these criteria will be described in more detail under separate headings.

Total Threshold Limit Concentrations (TTLCs)

Thirty-eight chemicals (or groups of chemicals) each have a numerical TTLC. If the concentration in the waste exceeds the TTLC for any of the 38 chemicals, the waste is categorized as hazardous, otherwise it is unregulated by DTSC. The existing TTLCs were designed to protect human health and the environment from adverse effects resulting from all exposure pathways other than exposure via ground water. The TTLC for each chemical

APPENDIX D: DTSC REPORT

was computed by multiplying the current STLC value by 100 or 10. TTLCs have no counterpart in federal regulation and are deemed important to retain. However, some serious problems exist with the current TTLC values:

Current TTLCs, being simply multiples of the STLCs are not necessarily relevant to non-ground water exposures. They are inflexible, in that they:

a. Require all wastes be either rigidly regulated or completely unregulated by DTSC

b. Cannot incorporate advances in technical information

c. Provide no guidance for decisions on specific waste streams (variances or reclassifications)

d. Do not provide a defined mechanism for regulating additional chemicals.

Proposed TTLCs were developed using a multi-step, risk-based approach:

1. Define potentially exposed humans for two waste management situations: 1) wastes managed as special wastes and disposed of in a landfill and 2) wastes unregulated by DTSC.

2. Identify pathways by which chemical in waste could reach humans for each scenario.

3. Develop a mathematical model to relate waste concentration to human health risk for each scenario.

4. Implement these models as spreadsheets to make computation easy and transparent.

5. Using the spreadsheets, compute a human-health-based TTLC for each scenario for each chemical.

6. For each chemical, determine if human health-based TTLCs protect other species. If not, compute a TTLC based on protection of non-human resources.

7. Evaluate risk-based TTLCs for science policy considerations.

The risk-based approach to deriving TTLCs addresses the four problems with the current TTLCs. The reader is encouraged to read Appendix 3 of Tab 3 for a general description of the derivation of TTLCs. This section is intended to be a "road map" directing the reader to the spreadsheet descriptions and showing where each fits into the overall process.

Human Health Risk Assessment Scenarios and Models

In order to develop a regulatory threshold, the fate and potential exposure scenarios and pathways for wastes that do not exceed the proposed threshold are evaluated. Thus, the scenarios for the upper TTLCs are associated with managed disposal in a municipal solid waste landfill meeting the specifications in RCRA subtitle D. Similarly, the scenarios for the exit-level TTLCs are associated with use of the waste as a soil amendment. There are many possible exposure scenarios, but DTSC has not identified any plausible scenarios in which exposures to toxic constituents in wastes is likely to exceed the exposures modeled using these scenarios. The scenarios and the models used to evaluate them are described below and summarized in a table following the description.

Upper TTLCs

The toxicologic risk scenarios evaluated for the upper TTLCs include workers at waste handling facilities and residents living adjacent to a waste handling facility. For the workers, a modified Preliminary Endangerment Assessment (PEA) model (see Tab 3, Appendices 3 & 4c) was used. This model was also used to estimate exposure of adjacent residents to inorganic constituents. Exposure of adjacent residents to organic constituents was estimated using the CalTOX model (Tab 3, Appendices 3 & 4a). Non-carcinogenic effects of lead were evaluated for on-site workers using the DTSC Lead Risk Assessment Spreadsheet (see Tab 3, Appendices 3 & 4b).

Exit-level TTLCs

The human health risk scenario evaluated for the lower (exit-level) TTLCs was a resident living on land on which the waste had been applied at the rate of 7000 kg/ha annually for 20 years and thoroughly mixed with the upper 15 cm of soil, such as might occur if the non-hazardous waste were used as a soil amendment and housing were subsequently built on the amended soil. The modified Preliminary Endangerment Assessment (PEA) model (see Tab 3, Appendices 3 & 4c) was used to estimate exposure of these residents to inorganic constituents. Exposure of the residents to organic constituents was estimated using the CalTOX model (see Tab 3, Appendices 3 & 4a). Non-carcinogenic effects of lead were evaluated for all scenarios using the DTSC Lead Risk Assessment Spreadsheet (see Tab 3, Appendices 3 & 4b). Potential effects on non-human receptors were also evaluated in establishing the exit-level TTLCs. The approaches used in this evaluation are described behind Tab 3, Appendices 3 & 4d.

The analyses used for the various scenarios are summarized in the following table:

Criterion	Scenario	Chemicals	Model	Further description
Upper TTLCs	Worker	Organics	modified PEA	Tabs3:Appendix 3 & 4c
Upper TTLCs	Worker	Inorganics	modified PEA	Tab 3:Appendix 3 & 4c
Upper TTLCs	Resident	Organics	CalTOX Landfill	Tab 3:Appendix 3 & 4a
Upper TTLCs	Resident	Inorganics	modified PEA	Tab 3:Appendix 3 & 4c
Upper TTLCs	Worker	Lead	Pb Spreadsheet	Tab 3:Appendix 3 & 4b
Exit-level TTLCs	Resident	Inorganics	modified PEA	Tab 3:Appendix 3 & 4c
Exit-level TTLCs	Resident	Organics	CalTOX Land Conversion	Tab 3:Appendix 3 & 4a
Exit-level TTLCs	Resident	Lead	Pb Spreadsheet	Tab 3:Appendix 3 & 4b
Exit-level TTLCs	Ecological	all	several	Tab 3:Appendix 3 & 4d

After the most protective risk based TTLC criteria were determined, those values were compared against 1) an estimated quatitation limit to determine if that concentration could be measured, and 2) the ambient concentration in soil. The derivation of these science policy values are described in Tabs 5a and 5b, respectively.

APPENDIX D: DTSC REPORT 157

Soluble or Extractable Regulatory Thresholds (SERTs)

The existing California waste classification system contains 36 Soluble Threshold Limit Concentrations (STLCs). To evaluate compliance with these criteria, a solid waste is extracted using a procedure called the Waste Extraction Test (WET), and the concentrations of the inorganic chemicals in the extract are compared with their corresponding STLCs. The extract is not normally analyzed for organic chemicals because it is mathematically impossible to exceed the STLC without also exceeding the TTLC. Liquid wastes are analyzed directly, without the extraction. The proposal would replace the WET with the RCRA Toxicity Characteristic Leaching Procedure (TCLP) and replace the 36 STLCs with 35 pairs of SERTs, corresponding to an upper level and an exit level standard for each regulated chemical. Silver would be dropped from the list and there would only be an upper-level SERT for polychlorinated biphenyls (PCBs). The approachs to developing SERTs and the associated extraction methodology are summarized in the table below. The methodology and equations for deriving the proposed SERTs are described in the waste classification concept paper (Tab 3, Appendix 2). The rationale and supporting documents for the proposal to replace the WET with the TCLP is found in Tab 6.

Criterion	Scenario	Pathway	Further description
Upper SERTs	Drinking water	ingestion	Tab 3:Appendix 3
Exit-level SERTs	Drinking water	ingestion	Tab 3:Appendix 3
Exit-level SERTs	Aquatic life	immersion	Tab 3:Appendix 3
Extraction procedure	leaching from a landfill		Tab 3:Appendix 2

2. Issues

The Department would like to highlight some key issues for your review: If the panel does not agree with any of the approaches the Department has used, we invite your suggestions for alternative approaches.

TTLC issues

Model Selection

- Is it appropriate to use a multimedia, multipathway risk assessment model (a variant of CalTOX) to establish concentration limits on organic chemical constituents in various classes of waste? (reference: Tabs 3, Appendix 3 and 4a)

- Is it appropriate to use a simple exposure model to establish concentration limits on inorganic chemical constituents in various classes of waste? (reference: Tab 3, Appendix 3 and 5c)

- Is it appropriate to use a lead uptake/blood lead model to establish concentration limits on lead in various classes of waste? (reference: Tab 3, Appendix 3 and 5b)

Scenario Selection

- Are the waste management worker and the nearby resident scenarios appropriate to represent populations potentially exposed to lower-tier hazardous wastes? The exposures of these populations would be used to establish a "bright line" to separate wastes which are subject to pre-disposal requirements and must be disposed of in RCRA subtitle C (hazardous waste) landfills from those which are subject to reduced pre-disposal requirements and can be safely disposed of in RCRA subtitle D (MSW) facilities? (reference: Tab 3, Appendix 3) The Department recognizes that some wastes may be disposed of in ways that have greater potential exposure and other wastes may be disposed of in ways that have less potential exposure. Soil is an example of a waste that has a high probability of being disposed of in a way that results in less dilution and therefore potentially greater exposure.

- All assessments of carcinogenic risk are based on estimated exposures to a composite person, having characteristics between those of a child and those of an adult. CalTOX also uses this approach for non-carcinogenic effects, while the PEA-based model and the Lead Risk Assessment Spreadsheet assesses exposure to children and adults separately for non-carcinogenic effects (the lowest value is selected for residential exposures while the adult value is used for workers). Which approach is appropriate for carcinogens? For non-carcinogens?

Pathways and Parameters

- Do the exposure pathways evaluated for workers and residents reasonably represent the exposure potential for these groups of people? (reference: Tabs 4a, 4b, and 4c)

- In the original submission it was assumed that 30% of a worker's skin is exposed. The 30% figure used in the assessment falls between worst case (shorts, no shirt, with about 60% of the skin exposed) and typical case (long pants, long- or short-sleeved shirt).

APPENDIX D: DTSC REPORT

The latter would involve about 13% exposure. The Department was able to obtain only anecdotal information about landfill workers' attire. Is 30% a reasonable mean value for fraction of skin exposed?

- Although daily cover of active disposal areas is required and only a limited area of a landfill may be active at any given time, no cover was assumed in modeling the landfill. This may be less important for vapors which are expected to escape through gas collection systems regardless of daily cover or capping. With adequate data, a distribution of uncovered area for California landfills and waste piles could theoretically be developed. However, DTSC used conservative modeling parameters partly because of data limitations and also because workers and residents may be exposed to vapors and particles from waste piles, which may not have daily cover.

- The degree of overestimation of exposure from a landfill source may be considerable for particulate pollutants because their release should be significantly reduced by daily cover or a cap. The degree of overestimation of exposure from a landfill source would likely be much less for vapors, because advective transport is likely to result in vapor emission, usually through a gas collection system, regardless of daily cover or permanent cap. However, only a fraction of the landfill would be producing gases at any given time.

- No dust dilution was assumed for PEA and Lead spreadsheet modeling, i.e. all respirable particulates are assumed to come from the waste. Monitoring data for California landfills and waste piles were sought, but most existing data are based on complaint follow-ups, and would be highly biased.

- The U.S. EPA has withdrawn the oral cancer potency factor for beryllium. Treating beryllium as a carcinogen only by the inhalation route would result in about a six-fold increase in the beryllium TTLC. Should beryllium be treated as an oral carcinogen?

Other Issues

- Is the proposed stepwise approach to ensuring that TTLCs based on human toxicity will protect non-human biota reasonable? (reference: Tab 3, Appendix 3 and 5d)

- Is the proposed use of two times the estimated quantitation level (EQL) in lieu of a TTLC when the calculated concentration is less than the EQL reasonable? (reference Tab 5a)

- Is the proposed approach to considering ambient levels of elements and compounds in setting TTLCs appropriate? (reference Tab 5b).

SERT issues

- SERTs are based on the lowest of (1) health-based concentrations calculated by DTSC, (2) California Maximum Contaminant Levels established by the Department of Health Services, or (3) U.S. EPA Ambient Water Criteria for the protection of aquatic life. Is this paradigm appropriate? (reference: Tab 3, Appendix 2)

- Is the dilution/attenuation factor of 100 for the SERTs appropriate? (reference: Tab 3, Appendix 2)

- Is the use of a factor to account for the retardation of leakage from the landfill due to the presence of a synthetic liner appropriate? (reference: Tab 3, Appendix 2) Was the liner protection factor calculated appropriately? (reference: Tab 7)
- The current STLCs and proposed SERTs are based on an assesment of the environmental effects of leachates from solid wastes in a landfill, but apply also to liquid wastes that are not disposed of in a landfill. Should the effects of released liquids be assessed separately? (reference: Tab 3, Appendix 2 & 6)
- Is laboratory extraction of toxic constituents from solid wastes with municipal solid waste leachates a reasonable standard against which to compare the performance of laboratory extraction protocols such as the WET, the TCLP, and the SPLP? Is the Department's conclusion that the WET is not a better estimator of leaching potential than the TCLP justified? (reference: Tab 6)
- We concluded that the extraction by municipal solid waste leachates of elements capable of forming oxyanions was not simulated by the WET or by the TCLP. Is this a reasonable conclusion? (reference Tab 3, Appendix 2 & 6)
- In the current proposal, ingestion of ground water is the only human exposure pathway evaluated. Does this approach seriously underestimate risk? (reference: Tab 3, Appendix 2)
- Is the use of the 90th percentile concentration of arsenic in monitored California drinking water supplies as a basis for the arsenic SERT appropriate? (reference Tab 3, Appendix 2)
- Is the proposed use of two times the estimated quantitation level (EQL) in lieu of a SERT when the calculated concentration is less than the EQL reasonable? (reference tab 5a)

Acute Toxicity Issues

- Is the Department's proposed approach to setting acute oral, dermal, and inhalation toxicity thresholds reasonable? (reference: Tab 3:Appendix 4)

APPENDIX D: DTSC REPORT

California Environmental Protection Agency
Department of Toxic Substances Control

Tab 3: Concept Paper

CALIFORNIA'S NON-RCRA WASTE CLASSIFICATION SYSTEM:

Analysis and Proposed Revisions

February 20, 1998

Authors: Jim Carlisle, D.V.M., M.Sc., Ned Butler, Ph.D., D.A.B.T., Kimi Klein, Ph.D., John Christopher, PhD, DABT, Bart Simmons, PhD.

Tab 4: Waste Classification Concept Paper

CALIFORNIA'S NON-RCRA WASTE CLASSIFICATION SYSTEM:
Analysis and Proposed Revisions

DESCRIPTION OF ISSUES

The Waste Classification Review

Waste classification review is one of the components of the Regulatory Structure Update project (RSU) of the Department of Toxic Substances Control (DTSC). The objective of the waste classification review is to ensure that California has a scientifically defensible method of classifying waste which protects human health and the environment without unnecessarily hindering sustainable growth and development. This is a multi-step process. The first task was to evaluate the protectiveness of the federal system under the Resource Conservation and Recovery Act (RCRA) with its listed hazardous wastes and the toxicity characteristic. The preliminary conclusion of this task is that the RCRA system is incomplete and does not, by itself, adequately protect human health and the environment in California (see draft concept paper for RSU task D-1). The RCRA system was intended to complement systems for classification and management of wastes in the states.

The Existing California System

The purpose of task D-3 is to evaluate the current California system against the four RSU criteria for change. Those aspects of the California system which remain relevant, appropriate, and scientifically defensible will be retained. Those that fail one or more of the RSU criteria will be modified or dropped.

Recommendation

Develop and adopt a revised California waste classification system establishing two tiers of regulated wastes, but retaining appropriate aspects of the current system.

Circumstances Necessitating a Decision

The finding that the RCRA system was incomplete as a basis for classifying wastes as hazardous in California (RSU task D-1) necessitated a determination of whether the present California waste classification regulations provide an adequate and tenable basis for identifying hazardous wastes in order to protect public health and the environment in California. In the twenty years since the major elements of this system were promulgated, the sciences of toxicology and risk assessment have advanced considerably. It is appropriate at this time to review the California system and update it as necessary.

APPENDIX D: DTSC REPORT   163

Existing Statutes or Regulations Related to the Issue

The Hazardous Waste Control Act of 1973 mandated that "the Department (then the Department of Health Services) shall "develop and adopt by regulation criteria and guidelines for the identification of hazardous wastes and extremely hazardous wastes" (Section 25141, Health and Safety Code). Section 25159 of the Health and Safety Code authorizes the Department to "adopt and revise when necessary regulations which will allow the State to receive and maintain authorization to administer a State hazardous waste program in lieu of the federal program..." Section 25159.5 of the Health and Safety Code "does not prohibit the Department from adopting standards and regulations which are more stringent or more extensive than federal regulations." Because of the broad authority provided in the original statutes, no new statutes will be required to revise existing regulations pertaining to criteria for identification of hazardous waste.

Existing State regulations describing the characteristic of ignitability are found in 22CCR 66261.21. No changes are proposed in this section because the State characteristic of ignitability is identical to the federal characteristic of ignitability.

Existing State regulations describing the characteristic of corrosivity are found in 22CCR 66261.22. Proposed changes to this section are the subject of RSU task C-11.

Existing State regulations describing the characteristic of reactivity are found in 22CCR 66261.23. No changes are proposed in this section because the State characteristic of reactivity is identical to the federal characteristic of reactivity.

Existing California regulations describing the characteristic of toxicity are found in 22CCR 66261.24. In addition to the RCRA toxicity characteristic, they include tables listing Soluble Threshold Limit Concentrations (STLCs) for 36 compounds. Wastes are analyzed using the Waste Extraction Test (WET, the subject of RSU task D-4), and exhibit the characteristic of toxicity if any of these chemicals are present at extractable concentrations exceeding the STLCs. The STLCs were established to protect ground water and are analogous to, though in some cases different from, the RCRA Toxicity Characteristic (TC) thresholds. RCRA TC thresholds have been established for 40 constituents. There are STLCs for 18 of these constituents. Thirteen of these 18 have the same numerical value. Of the five which have different values, California's standard is higher (less restrictive) for four (chlordane, heptachlor, TCE and trivalent chromium), and lower for one (pentachlorophenol). Since the current test methodology is different, the California standard can be more restrictive, even when the numerical standards are the same.

California regulations in 22CCR 66261.24 include lists of Total Threshold Limit Concentrations (TTLCs) for 38 compounds. The TTLCs were established to protect against human health and environmental impacts resulting from improper management of toxic wastes. Exposure of humans or other animals by ingestion, inhalation, and dermal absorption of toxic compounds was not analyzed directly, but thresholds were established based on a multiple, usually 100, of the STLCs or were established independently based on observed deaths or injuries to humans or livestock.

California regulations also identify wastes as hazardous if they are acutely toxic to mammals or fish, or if they contain any of sixteen listed carcinogenic compounds at a concentration exceeding 10 ppm. Acute oral, dermal, and inhalation toxicity limits were established by multiplying the "highly hazardous" levels recommended by NIOSH (1974) by a safety factor of 100. The fish toxicity threshold is the value used by the federal government to identify commercial substances which may pose a hazard if spilled or discharged into an aquatic environment (40 CFR 116). The carcinogen list is from the Department of Industrial Relations and was produced pursuant to the California Workplace Right to Know Law.

Existing regulations identify as hazardous any waste that has been shown through experience or testing to pose a hazard to human health or the environment for a variety of reasons. The regulations also include a list of presumptively hazardous compounds.

Objectives to be Achieved Through Decision

The proposed action supports the mission of the Department, which is to protect public health and the environment without unnecessarily hindering sustainable growth and development. The recommended changes emphasize (1) consistency with federal standards when possible, (2) maintenance or enhancement of the level of protection when such protection is justified by the waste's toxicity and its likelihood of exposure to humans and non-human biota, (3) reduction of unnecessary costs associated with management of wastes as hazardous when the incremental benefit is minimal, (4) minimization of unnecessary use of limited hazardous waste landfill capacity, and (5) consistency with the four general RSU criteria for decision-making which are:

1. Are the environmental and public health circumstances and/or lack of federal/state regulatory oversight that led to the establishment of the requirement still in existence?

2. Are there other regulatory requirements or agencies which adequately cover the concerns addressed by the requirement?

3. Is the requirement a rational use of environmental protection resources given the potential risks and their probability and significance?

4. Is there a regulatory alternative that would achieve the desired human health and safety or environmental benefit in a more cost-effective and flexible manner?

Alternatives Available and Recommended Alternative

A. Subsection 66261.24(a)(2)(A&B)

Subsection 66261.24(a)(2)(A&B) contains a list of Total Threshold Limit Concentrations (TTLCs) for specific contaminants in mg/kg waste. If a solid waste contains any of the listed contaminants at a concentration exceeding its TTLC, the waste is a hazardous. With respect to the Total Threshold Limit Concentrations, the options considered were:

1. Repeal the TTLCs in 22CCR Subsection 66261.24 (a)(2)(A&B)

   Pro: This alternative would have the advantage of being consistent with RCRA, eliminating the need to perform total extraction on wastes, reducing regulatory burden, and reducing administrative costs.

   Con: This alternative would provide for no limits on total concentrations of hazardous constituents in wastes, thereby reducing protectiveness below that mandated by State statutes. The RCRA toxicity characteristic does not address exposure to toxic constituents via media other than ground water. Constituents that do not exceed the federal toxicity

APPENDIX D: DTSC REPORT 165

characteristic thresholds can contaminate other environmental media such as air, soil, and sediments and therby harm public health and the environment.

2. Retain the TTLCs 22CCR Subsection 66261.24(a)(2)(A&B)in their current form.

   Pro: This alternative would partially protect against harm to public health and the environment through airborne dispersion or particulate dispersion via surface water or by direct exposure to the wastes

   Con: The current TTLCs are not based on a consistent multipathway risk analysis. Several are simple multiples of the STLCs. Some do not reflect the most current toxicological data. The all-or-none approach results in over-regulating some wastes and under-regulating others. This alternative would retain a regulatory system more stringent than RCRA.

3. Using a multipathway analytical approach, develop and adopt two tiers of revised TTLCs for constituents for which TTLCs currently exist. Develop TTLCs for additional constituents as permitted by resources and data availability.

   Pro: This alternative would address the identified shortcomings of the RCRA and current California systems.

   Con: This alternative would retain a regulatory system more stringent than RCRA. A two-tier system adds complexity to classification of wastes and is expected to increase confusion and need for classification assistance and training. It may increase the volume of wastes disposed in some municipal landfills and could decrease incentives to recycle.

4. Using a multipathway analytical approach, develop and adopt revised TTLCs for constituents for which TTLCs currently exist. Develop TTLCs for additional constituents as permitted by resources and data availability.

   Pro: This alternative would address the identified shortcomings of the RCRA and current California systems.

   Con: This alternative would retain a regulatory system more stringent than RCRA. Since wastes would be either in the DTSC regulatory system or out, without a middle ground for low-risk wastes, some low-risk wastes would be under-regulated, others would be over-regulated.

Recommended Alternative: Alternative 3

Basis of selection:

Criterion 1. The environmental and public health circumstances that led to the regulatory thresholds (TTLCs) in 22CCR Subsection 66261.24(a)(2)(A&B) have changed. In particular, knowledge of toxicological effects, exposure pathways, and the use of mathematical models have advanced considerably since the time the current regulations were promulgated.

Criterion 2. There are no other state lists and no state agencies authorized to create lists of threshold levels of chemicals in wastes for the purpose of defining those wastes as hazardous. The impact of potential exposure by multiple pathways is not considered by the RCRA toxicity characteristic (although multiple pathways were considered in some listings).

Criterion 3. Retaining a list of Total Threshold Limit Concentrations to identify a waste as hazardous is necessary to protect public health and the environment in California. Such a list provides the public with clear guidance in classifiying a waste. Revising the TTLCs using up-to-date multimedia, multipathway risk assessment for both humans and ecosystems is necessary to retain scientific credibility. The proposed revised TTLCs meet this criterion by employing methodology that evaluates risks with respect to probability and significance.

Criterion 4. A more cost-effective and flexible regulatory alternative that met the protectiveness standard was not identified. By establishing two subclasses of hazardous wastes using two tiers of TTLCs as toxicity criteria, flexibility in handling and disposing of hazardous waste is integrated into the HWMP. By attaching different requirements to those wastes that exceed the levels, cost-effective management of hazardous waste is achieved.

B. Subsection 66261.24(a)(2)(A&B)

Subsection 66261.24(a)(2)(A&B) contains a list of Soluble Threshold Limit Concentrations (STLCs) expressed in mg/l for specific constituents in liquid waste or the extract from the Waste Extraction Test (WET) performed on a solid waste. DTSC's regulatory program with respect to threats to the waters of the State has RCRA and non-RCRA components. RCRA regulations specify Toxicity Characteristic Regulatory Limits (RLs) for 40 constituents and California has STLCs for 36 compounds. Nineteen compounds have both STLCs and RCRA RLs. With respect to the Soluble Threshold Limit Concentrations, the options considered were:

1. Repeal the STLCs in 22CCR Subsection 66261.24 (a)(2)(A&B)

    Pro: This alternative would have the advantage of being consistent with RCRA, reducing regulatory burden, and reducing administrative costs.

    Con: This alternative would not provide soluble or extractable concentrations of hazardous constituents in wastes for which there are no RCRA regulatory limits and would not allow for stricter State limits to fully account for carcinogenic effects and/or deficiencies in the TCLP test method for constituents for which there are RCRA regulatory limits.

2. Retain the STLCs in 22CCR Subsection 66261.24(a)(2)(A&B)in their current form.

    Pro: This alternative would partially protect against harm to public health and the environment from exposure to soluble or extractable concentrations of hazardous constituents in wastes for which there are no RCRA regulatory levels.

    Con: This alternative would not allow for updating the STLCs to reflect the most current toxicology data and to account for deficiencies in the test methodology. The all-or-none approach results in over-regulating some wastes and under-regulating others. This alternative would retain inconsistency with RCRA and other states.

3. Revise 22CCR Subsection 66261.24(a)(2)(A&B) to establish two tiers of Soluble or Extractible Regulatory Thresholds (SERTs), which would replace the current STLCs. Develop SERTs for additional constituents as permitted by resources and data availability.

    Pro: This alternative would address the identified shortcomings of the RCRA and current California systems.

    Con: This alternative would retain inconsistency with RCRA and other states.

4. Revise 22CCR Subsection 66261.24(a)(2)(A&B) to establish risk-based STLCs. Develop STLCs for additional constituents as permitted by resources and data availability.

    Pro: This alternative would address the identified shortcomings of the RCRA and current California systems.

    Con: This alternative would retain inconsistency with RCRA and other states. Since wastes would be either in the DTSC regulatory system or out, without a middle ground for low-risk wastes, some low-risk wastes would be under-regulated, others would be over-regulated.

APPENDIX D: DTSC REPORT

Recommended Alternative: Alternative 3

Basis of selection:

Criterion 1. The circumstances that led to the regulatory thresholds (STLCs) in 22CCR Subsection 66261.24(a)(2)(A&B) have changed. In particular knowledge of toxicological effects on humans and aquatic ecosystems have advanced considerably since the time the current regulations were promulgated. A list of Soluble or Extractible Regulatory Thresholds (SERTs, formerly Soluble Threshold Limit Concentrations, STLCs) for specific chemicals as part of a comprehensive methodology to classify wastes continues to be necessary to protect the environmental and public health in California, because such a list provides the public with clear guidance regarding the categorization of a waste as hazardous. Revising the SERTs using the most recent toxicological data to evaluate the impacts of these chemicals on aquatic systems and on human health will maintain the scientific credibility of these standards.

Criterion 2. There are no other state lists and no state agencies authorized to create lists of threshold levels of chemicals in wastes for the purpose of defining those wastes as hazardous. Protection of the waters of the State requires a coordinated approach between DTSC and the State Water Resources Control Board and the Regional Water Quality Control Boards. These Boards regulate discharge of wastes to land. DTSC currently classifies wastes as hazardous or extremely hazardous. Land disposal of hazardous wastes generally must be in a Class 1 landfill. Wastes that are not classified by DTSC as hazardous may be classified by the Boards as "designated", "non-hazardous solid", or "inert". These classifications require that land disposal be in at least class II, class III, or unclassified landfills, respectively. DTSC's hazardous waste management program and those of the Water Boards must mesh in a seamless manner. DTSC will facilitate this by making it clear that wastes that are not regulated by DTSC may be subject to other regulatory requirements.

Criterion 3. The proposed revised SERTs are derived using methodology that evaluates risks in regard to probability and significance, and, therefore, meets this criterion.

Criterion 4. A more cost-effective and flexible regulatory alternative that met the protectiveness standard was not identified. By establishing two subclasses of hazardous wastes using two tiers of SERTs as toxicity criteria, flexibility in handling and disposing of hazardous waste is integrated into the HWMP. By attaching different requirements to those wastes that exceed the levels, cost-effective management of hazardous waste is achieved.

C. Subsection 66261.24 (a)(3)

Subsection 66261.24 (a)(3) contains a lower limit on the acute oral median lethal dose (LD_{50}, the dose of the waste that is fatal to half of the animals exposed) of 5000 mg/kg. Existing statute lowers that threshold to 2500 mg/kg. With respect to the acute oral LD_{50}, the options considered were:

1. Repeal 22CCR Subsection 66261.24 (a)(3)

    Pro: This alternative would have the advantage of being consistent with RCRA, reducing regulatory burden, and reducing administrative costs.

    Con: This alternative would not permit identification of wastes as hazardous based on their acute oral toxicity. This would reduce protectiveness below that mandated by state statutes because there may be waste constituents or combinations of waste constituents that may be acutely toxic for which chronic toxicity benchmarks have not been established.

2. Retain 22CCR Subsection 66261.24(a)(3) in its current form.

Pro: This alternative would protect against harm to public health from oral exposure to acutely toxic constituents in wastes for which there are no established chronic toxicity benchmarks.

Con: This alternative would continue the all-or-none approach which results in over-regulating some wastes and under-regulating others, and would retain a classification that differs from the federal system and other states' systems.

3. Develop and adopt a revised 22CCR Subsection 66261.24(a)(3) establishing two acute oral toxicity limits corresponding to the two tiers of regulated wastes.

Pro: This alternative would address the identified shortcomings of the RCRA and current California systems.

Con: This alternative would retain inconsistency with RCRA and other states.

4. Develop and adopt a revised 22CCR Subsection 66261.24(a)(3) establishing a risk-based acute oral toxicity limit.

Pro: This alternative would address the identified shortcomings of the RCRA and current California systems.

Con: This alternative would continue the all-or-none approach which results in over-regulating some wastes and under-regulating others, and would retain a classification that differs from the federal system and other states' systems.

Recommended Alternative: Alternative 3

Basis of selection:

See below under Subsection 66261.24 (a) (5).

D. Subsection 66261.24 (a)(4)

E. Subsection 66261.24 (a)(4) contains a lower limit on the acute dermal median lethal dose (dermal LD_{50}) of 4300 mg/kg. With respect to the acute dermal LD_{50}, the options considered were:

1. Repeal 22CCR Subsection 66261.24 (a)(4)

Pro: This alternative would have the advantage of being consistent with RCRA, reducing regulatory burden, and reducing administrative costs.

Con: This alternative would not permit identification of wastes as hazardous based on their acute dermal toxicity. This would reduce protectiveness below that mandated by state statutes because there may be waste constituents or combinations of waste constituents that may be acutely toxic for which chronic toxicity benchmarks have not been established.

2. Retain 22CCR Subsection 66261.24(a)(4) in its current form.

Pro: This alternative would protect against harm to public health from dermal exposure to acutely toxic constituents in wastes for which there are no established chronic toxicity benchmarks. It is important to separately evaluate dermal toxicity because chemicals that are well absorbed through the skin may present unique hazards.

Con: This alternative would continue the all-or-none approach which results in over-regulating some wastes and under-regulating others. This alternative would retain inconsistency with RCRA and other states.

APPENDIX D: DTSC REPORT

3. Develop and adopt a revised 22CCR Subsection 66261.24(a)(4) establishing two acute dermal LD_{50} limits corresponding to the two tiers of regulated wastes.

    Pro: This alternative would address the identified shortcomings of the RCRA and current California systems.

    Con: This alternative would retain inconsistency with RCRA and other states.

4. Develop and adopt a revised 22CCR Subsection 66261.24(a)(4) establishing a risk-based acute oral toxicity limit.

    Pro: This alternative would address the identified shortcomings of the RCRA and current California systems.

    Con: This alternative would continue the all-or-none approach which results in over-regulating some wastes and under-regulating others, and would retain a classification that differs from the federal system and other states' systems.

Recommended Alternative: Alternative 3

Basis for Selection:

See below under Subsection 66261.24 (a)(5).

E. Subsection 66261.24 (a)(5)

Subsection 66261.24 (a)(5) contains a lower limit on the acute inhalation median lethal concentration (LC_{50}, the concentration of the waste that is fatal to half of the animals exposed) of 10,000 ppm as a gas or vapor. With respect to the acute inhalation LC_{50}, the options considered were:

1. Repeal 22CCR Subsection 66261.24 (a)(5)

    Pro: This alternative would have the advantage of being consistent with RCRA, reducing regulatory burden, and reducing administrative costs.

    Con: This alternative would not permit identification of wastes as hazardous based on their acute inhalation toxicity. This would reduce protectiveness below that mandated by state statutes because there may be waste constituents or combinations of waste constituents that may be acutely toxic for which chronic toxicity benchmarks have not been established.

2. Retain 22CCR Subsection 66261.24(a)(5) in its current form.

    Pro: This alternative would protect against harm to public health from inhalation exposure to acutely toxic constituents in wastes for which there are no established chronic toxicity benchmarks.

    Con: This alternative would continue the all-or-none approach which results in over-regulating some wastes and under-regulating others. This alternative would retain inconsistency with RCRA and other states.

3. Develop and adopt a revised 22CCR Subsection 66261.24(a)(5) establishing two acute inhalation LC_{50} limits corresponding to the two tiers of regulated wastes, allowing calculated head-space vapor concentrations, and adjusting for percent respirable particles.

    Pro: This alternative would address the identified shortcomings of the RCRA and current California systems.

Con: This alternative would retain inconsistency with RCRA and other states and would add complexity to the regulations.

4. Develop and adopt a revised 22CCR Subsection 66261.24(a)(5) establishing acute inhalation LC_{50} limits, allowing calculated head-space vapor concentrations, and adjusting for percent respirable particles.

Pro: This alternative would address the identified shortcomings of the RCRA and current California systems.

Con: This alternative would retain inconsistency with RCRA and other states and would add complexity to the regulations. It would also continue the all-or-none approach which results in over-regulating some wastes and under-regulating others

Recommended Alternative: Alternative 3

Basis for selection of alternatives presented for subsections 66261.24 (a) (3), (4), and (5):

Criterion 1. The circumstances leading to the adoption of acute toxicity criteria have not changed. Consideration of acute oral, dermal, and inhalation toxicity remain necessary in classifying waste in order to protect the public health. Acute toxicity is still not considered in the RCRA Toxicity Characteristic and, thus, represents a gap in federal regulatory oversight.

Criterion 2. There are no other state lists and no state agencies authorized to create criteria for acute toxicity of chemicals in wastes for the purpose of defining those wastes as hazardous. The RCRA program does not include acute toxicity limits in the Toxicity Characteristic.

Criterion 3. Specific waste constituents or combinations of waste constituents may be acutely toxic if accidently ingested, spilled on the skin, or inhaled, and may not have chronic toxicity benchmarks. It is reasonable to use this regulatory framework to protect human health from such exposure.

Criterion 4. A more cost-effective and flexible alternative to the proposal has not been identified. By establishing two subclasses of hazardous wastes using two tiers of acute lethal dose (LD_{50}) or acute lethal concentration (LC_{50}) as toxicity criteria, flexibility in handling and disposing of hazardous waste based on relative toxicity is integrated into the HWMP. By attaching different requirements to those wastes that exceed the levels, cost-effective management of hazardous waste is achieved.

F. Subsection 66261.24 (a)(6)

Subsection 66261.24 (a)(6) contains a lower limit on the 96-hour aqueous median lethal concentration (LC_{50}, the concentration of the waste that is fatal to half of the fish exposed) of 500 mg/l. With respect to the aquatic LC_{50}, the options considered were:

1. Repeal 22CCR Subsection 66261.24(a)(6)

   Pro: This alternative would have the advantage of being consistent with RCRA, reducing regulatory burden, and reducing administrative costs.

   Con: This alternative would not permit identification of wastes as hazardous based on their toxicity to fish. This would reduce protectiveness below that mandated by state statutes because there may be waste constituents or combinations of waste constituents that may be acutely toxic to fish for which chronic toxicity benchmarks have not been established.

2. Retain 22CCR Subsection 66261.24(a)(6) in its current form.

APPENDIX D: DTSC REPORT

Pro: This alternative would protect against harm to aquatic organisms from exposure to substances which are acutely toxic to fish but are not hazardous by any other criterion.

Con: This alternative would retain inconsistency with RCRA and other states and could over-regulate some wastes.

3. Develop and adopt a revised 22CCR Subsection 66261.24(a)(6) establishing two acute aquatic LC_{50} limits corresponding to the two tiers of regulated wastes.

Pro: This alternative would be consistent with the proposed two-tier standards for the other toxicity criteria.

Con: An upper tier of fish LC_{50} limits (creating an upper tier of hazardous wastes based on fish toxicity) would over-regulate some wastes because Class 1 containment is not necessary to mitigate the hazard to fish posed by direct exposure. This alternative would retain inconsistency with RCRA and other states.

4. Retain 22CCR Subsection 66261.24(a)(6) but use the fish bioassay results only to bring wastes into the lower tier of hazardous wastes.

Pro: This alternative would protect against harm to aquatic organisms from exposure to substances which are acutely toxic to fish but are not hazardous by any other criterion without imposing unnecessarily stringent requirements on some wastes. Class 1 containment is not necessary to mitigate the hazard to fish posed by direct exposure.

Con: This alternative would retain inconsistency with RCRA and other states.

Recommended Alternative: Alternative 4

Basis for Selection:

Criterion 1. The circumstances leading to the adoption of the aquatic toxicity criterion have not changed. The people of California still expect regulatory agencies to protect the environment. Aquatic toxicity is generally not covered by the RCRA program and, thus, represents a lack of federal oversight.

Criterion 2. The State Water Resources Control Board has regulatory oversight over waters of the State. However, the Board does not have the authority to define wastes as hazardous. Without that ability, the the Board may not have sufficient authority to regulate wastes in a manner that fully protects aquatic resources.

Criterion 3. Protecting aquatic life by requiring the performance of a fish bioassay is a rational use of resources, as this is the only test required that directly measures the impact of a waste constituent on aquatic life, and it is the only test that directly measures the integrated toxicity of a waste as a whole.

Criterion 4. The requirement that wastes failing this test need only be managed as special wastes provides the regulated public with the most cost-effective alternative and maximum flexibility, given the mandate of the Department to protect the environment.

G. Subsection 66261.24 (a)(7)

Subsection 66261.24 (a)(7) contains a list of sixteen carcinogenic compounds which, if present at more than 10 ppm in the waste, render it hazardous. The options for this subsection were:

1. Repeal 22CCR Subsection 66261.24(a)(7)

Pro: This alternative would have the advantage of being consistent with RCRA, reducing regulatory burden, and reducing administrative costs. Most of the listed carcinogens are RCRA-listed hazardous wastes.

Con: This alternative could, by itself result in partially deregulating several carcinogens. RCRA-listed hazardous wastes are only hazardous to the extent that they meet the RCRA definitions.

2. Retain 22CCR Subsection 66261.24(a)(7) in its current form.

Pro: This alternative would partially protect against harm to public health from exposure to the listed carcinogens.

Con: This alternative would retain the single threshold which does not recognize the varying carcinogenic potency of the listed compounds. This alternative would continue the all-or-none approach which results in over-regulating some wastes and under-regulating others, and would retain inconsistency with RCRA and other states.

3. Repeal 22CCR Subsection 66261.24(a)(7). Add vinyl chloride to the TTLC and SERT lists and add the balance of the listed carcinogens to Appendix X to section 66261. Classify future identified carcinogens via TTLCs and SERLs.

Pro: This alternative would provide for a consistent methodology for assessing risk of carcinogenic effects and with the proposed two-tier standards for the other toxicity criteria and would be consistent with the proposed two-tier standards for the other toxicity criteria. There is no scientific reasons to consider carcinogens as a separate class of hazardous wastes. This alternative would allow generators of the formerly listed carcinogens to demonstrate that their wastes are not hazardous. Several of the proposed TTLCs are based on carcinogenicity.

Con: This alternative would retain inconsistency with RCRA and other states and would reduce the control over the fifteen chemicals moved to Appendix X, because they would no longer have regulatory thresholds.

4. Retain 22CCR Subsection 66261.24(a)(7). Remove vinyl chloride from the carcinogen list, but add it to the TTLC and SERT lists. Classify future identified carcinogens via TTLCs and SERTs.

Pro: This alternative would provide risk-based thresholds for vinyl chloride, while not decreasing protection for the other listed carcinogens.

Con: This alternative would retain inconsistency with RCRA and other states and would retain the non-risk-based thresholds for fifteen carcinogens.

Recommended Alternative: Alternative 4

Basis for selection:

Criterion 1. The compounds listed in existing regulation were originally identified as carcinogenic by the Occupational Safety and Health Administration (OSHA) in 1977. Since that time, much research has been done on carcinogenic compounds, many additional compounds have been found to be carcinogenic, and the majority of compounds on the list are no longer in general commerce. Consideration of carcinogenicity is now an integral component of assessing risk of chronic exposure to chemicals. Public health circumstances have not changed with regard to the requirement of regulatory agencies to protect against undue exposure to carcinogenic compounds. Therefore, although the current list is very much out-of-date, consideration of carcinogenicity in the evaluation of wastes is still necessary.

APPENDIX D: DTSC REPORT 173

Criterion 2. Although many of the chemicals on this current list are listed on the RCRA U or P list, they may not be captured as RCRA-listed hazardous wastes because of definitional limitations.

Criterion 3. Since research has shown that chemicals exhibit varying carcinogenic potency, regulating sever carcinogenic compound at a single concentration limit would not result in equal protection from contracting cancer. The risk assessment methods used to establish SERT and TTLC values include carcinogenicity, thus, carcinogenic chemicals would be evaluated and listed in Subsection 66261.24 (a)(2). Vinyl chloride will be added to the SERT and TTLC list on the basis of its carcinogenicity.

Criterion 4. Repealing this list may be cost-effective, but may not meet DTSC's statutory mandates.

H. Subsection 66261.24 (a)(8)

Subsection 66261.24 (a)(8) identifies as hazardous any waste that has been shown through experience or testing to pose a hazard to human health or the environment for any of several reasons. With respect to this so-called new threats category, the options considered were:

1. Repeal 22CCR Subsection 66261.24(a)(8)

   Pro: This alternative would have the advantage of being consistent with RCRA,(the Federal system lacks the ability to administratively identify a waste as hazardous on the basis of new knowledge), potentially reducing regulatory burden, and reducing administrative costs.

   Con: This alternative would remove DTSC's ability administratively identify a waste as hazardous on the basis of new knowledge. Without this ability, continued human exposure or environmental harm could occur while DTSC goes through the rulemaking process to regulate a new or newly identified hazard.

2. Retain 22CCR Subsection 66261.24(a)(8) in its current form.

   Pro: This alternative would allow the Department to administratively identify a waste as hazardous on the basis of new knowledge. This ability is important in order to prevent continued human exposure or environmental harm whileDTSC goes through the rulemaking process to regulate a new or newly identified hazard.

   Con: This alternative could potentially increase regulatory burden and administrative costs.

3. Revise 22CCR Subsection 66261.24(a)(8).

   Pro: None identified

   Con: No needed revisions to this section were identified.

Recommended Alternative: Alternative 2

Basis for Selection:

Criterion 1. The circumstances leading to the adoption of the "new threats" criterion have not changed. Protection of the environment and public health requires that this toxicity characteristic be retained to protect against undue exposure to future wastes that are found to be hazardous to public health or the environment.

Criterion 2. There are no other state agencies authorized to create criteria to classify chemicals in wastes for the purpose of defining those wastes as hazardous. The RCRA program does not include carcinogenicity, toxicity, bioaccumulative or persistent properties as part of the toxicity characteristic, and cannot administratively list wastes as hazardous.

Criterion 3. The subsection represents a rational use of resources because of its general wording that recognizes that future experience and testing will likely uncover chemicals that should be identified as hazardous wastes.

Criterion 4. The subsection provides flexibility to the Department in identifying specific wastes as hazardous.

Implementation of the Recommended Alternatives

A. Modifications to Section 66261.24(a)(2): SERTs

DTSC proposes to replace the STLCs with two-tiered regulatory limits called Soluble or Extractible Regulatory Thresholds (SERTs). Wastes containing extractible concentrations of regulated constituents exceeding the upper SERT would be classified as non-RCRA hazardous wastes. Wastes containing extractible concentrations of regulated constituents exceeding only the lower (exit-level) SERT would be classified as Special Wastes, assuming that the waste is not hazardous by any other criterion. Special Wastes are a subset of hazardous wastes to which reduced regulatory requirements would apply. Table 1 presents proposed upper and lower SERTs along with current STLCs. If the SERT is >1/20 of the TTLC, the SERT applies only to liquid wastes. Derivation of proposed SERTs is presented in Appendix 2.

APPENDIX D: DTSC REPORT

Table 1: Proposed SERTs vs. current standards (mg/l)

	Proposed	based on	Proposed lower	based on	current	current RCRA
Aldrin	0.4	EQL e	0.006	TTLC	0.14	none
Chlordane	0.008	AWQC a	0.0007	EQL e	0.25	0.03
DDT & metabolites	0.005	EQL e	0.0005	EQL e	0.1	none
2,4-D	100	MCL c	7	MCL c	10	10
Dieldrin	0.009	EQL e	0.0009	EQL e	0.8	none
Endrin	0.008	EQL e	0.0008	EQL e	0.02	0.02
Heptachlor	0.008	EQL e	0.0008	EQL e	0.47	0.008
Kepone	0.4	HBL b	0.04	EQL e	2.1	none
Lindane	0.1	AWQC a	0.008	AWQC a	0.4	0.4
Methoxychlor	0.05	AWQC a	0.003	AWQC a	10	10
Mirex	0.003	EQL e	0.0003	EQL e	2.1	none
Pentachlorophenol	2	MCL c	0.1	MCL c	1.7	100
PCB	5 d	existing	none d		5	none
PCDD/PCDF	none		none		0.001	none
Toxaphene	0.02	EQL e	0.002	EQL e	0.5	0.5
TCE	9	MCL c	0.5	MCL c	204	0.5
2,4,5-TP (Silvex)	90	MCL c	5	MCL c	1	1
vinyl chloride	0.9	MCL c	0.05	MCL c	none	0.2
organic lead	20	EQL e	2	EQL e		
Antimony h	10	EQL e	1	EQL e	15	none
Arsenic h	59	HBL b	1	ambient	5	5
Barium	2000	MCL c	100	MCL c	100	100
Beryllium	1	EQL e	0.1	EQL e	0.75	none
Cadmium	2	AWQC a	0.1	AWQC a	1	1
Hexavalent chromium	20	HBL b	0.2	HBL b	5	none
Total chromium	90	MCL c	5	MCL c	560	5
Cobalt	4000	HBL b	200	HBL b	80	none
Copper	20	AWQC a	1	AWQC a	25	none
Fluoride	4000	HBL b	200	HBL b	180	none
Mercury	0.04	EQL e	0.004	EQL e	0.2	0.2
Molybdenum h	300	HBL b	10	HBL b	350	none
Nickel	200	MCL c	10	MCL c	20	none
Lead (inorganic)	5	AWQC a	0.3	AWQC a	5	5
Selenium h	9	AWQC a	0.5	AWQC a	1	1
Thallium	4	MCL	0.2	MCL	7	none
Vanadium h	500	HBL b	20	HBL b	24	none
Zinc	200	AWQC a	10	AWQC a	250	none

a. Based on federal Ambient Water Quality Criteria @ 100 ppm hardness or a pH of 6.5 and the longest available exposure, and a dilution/attenuation factor of 100 (the initial value of 0.0002 for toxaphene was raised to the PQL).

b. DTSC health-based concentration multiplied by 100
c. The California Maximum Contaminant Level (MCL) multiplied by 100
d. No new standard is proposed for PCBs
e. Based on the estimated quantification limit.
f. If the proposed value exceeds the RCRA RL, then the proposed value would apply only to federally exempt wastes.
g. The calculated value was 4, which is not sufficiently different from the current STLC to warrent a change.
h. Because of limitations in the TCLP, SERTs for this chemical apply only to liquid wastes

B. Modifications to Section 66261.24(a)(2): TTLCs

California currently has Total Threshold Limit Concentrations (TTLCs) for 38 substances. The Department proposes to replace existing TTLCs with two revised TTLC values - an upper and a lower (exit-level) TTLC - for each of 34 constituents. Silver is droped from the list because it is not believed to represent a significant threat. The current TTLC for PCBs is being retain pending anticipated changes in analytical methods and toxicological benchmarks. No upper TTLCs are proposed for zinc and total chromium because the calculated values for these standards are at or near 1 million ppm. Proposed TTLCs, along with the existing TTLCs are shown in Table 2:

APPENDIX D: DTSC REPORT

Table 2: Proposed TTLCs and Current TTLCs (mg/kg)

Chemical	Upper TTLC[a]	Basis[d]	Lower TTLC[a]	Basis[d]	Existing TTLC
Aldrin	0.7	EQL	0.05	EQL	1.4
Chlordane	1	HBL	0.06	HBL	2.5
DDT & congeners	1	HBL	0.3	HBL	1
2,4 D	1500	HBL	none[f]		100
Dieldrin	0.9	EQL	0.06	EQL	8
PCDD/PCDF	0.0002	ambient	none[f]		.01
Endrin	70	HBL	0.1	HBL	0.2
Heptachlor	0.8	EQL	none[f]		4.7
Kepone	40	EQL	3	EQL	21
Lead (organic)	100	EQL	10	EQL	13
Lindane	30	HBL	5	HBL	4
Methoxychlor	2000	HBL	100	HBL	100
Mirex	0.9	HBL	0.04	HBL	21
Pentachlorophenol	500	HBL	none[f]		17
Polychlorinated biphenyls (PCBs)	50	existing[b]	none		50
Tricholoethylene (TCE)	20	HBL	none[f]		2040
Toxaphene	2	EQL	0.1	EQL	10
Silvex	1000	HBL	none[f]		5
Vinyl chloride	1	EQL	0.2	EQL	.01
Antimony	700	HBL	none[f]		500
Arsenic	40	HBL	none[f]		500
Asbestos	none	na	1% (by wt.)	existing[b]	1% (by wt.)
Barium (excluding barite)	100,000	HBL	none[f]		10,000
Beryllium	20	HBL	none[f]		75
Cadmium	150	HBL	60	HBL	100
Hexavalent Chromium	5	HBL	none[f]		500
Total Chromium	none [h]	HBL	none [h]	HBL	2500
Cobalt	15,000	HBL	none[f]		8000
Copper	70,000	HBL	none[f]		2,500
Fluoride	100,000	HBL	none[f]		18,000
Lead	6000	HBL	1000	HBL	1,000
Mercury[c]	500	HBL	7	ambient	20
Molybdenum	9000	HBL	none[f]		3,500
Nickel	3000	HBL	none[f]		2,000
Selenium	30	STLC	none		100
Silver	none (see text)		none (see text)		500
Thallium	150	HBL	none[f]		700
Vanadium	1000	STLC	none		2,400
Zinc	none [g]	na	none [g]	na	5,000

7/2/98

b. The TTLCs for asbestos and PCBs were not reevaluated and will remain unchanged.
c. Wastes containing elemental mercury at any concentration are hazardous.
d. EQL = estimated quantitation limit; HBL = health-based limit; ambient - see text
f. The calculated lower TTLC was not significantly lower than the upper TTLC see text
g. No regulatory limit proposed - calculated value is at least 400,000 mg/kg.

Under the proposal, as under current regulation, wastes would be analyzed for the chemical constituents for which TTLC values exist. As in current regulation, measured total concentrations in a waste would be compared with listed TTLC values to determine how to classify the waste. If the total concentration measured in the waste exceeds the upper TTLC for any of the chemicals in the table, then the waste is a fully regulated hazardous waste. If the measured concentration in the waste is below the lower TTLC for all of the chemicals in those tables and passes all the other criteria, then the waste is not subject to regulation by DTSC. If the concentration of any constituent measured in the waste is between the upper and lower TTLC values, and it is not fully hazardous by any other criterion, then the waste will be classified as a California special waste. An explanation of the derivation of the proposed TTLCs is in Appendix 3. Appendix 1 contains brief toxicological profiles and basis for regulatory limits for the constituents for which DTSC is developing regulatory thresholds.

Regulation of Wastes Containing Elemental Metals

The Department currently only regulates elemental metals in wastes "...if the substances are in a friable, powdered, or finely divided state [22CCR, §66261.24(a)(2)(A)]." The current working definition for powdered or finely divided is material which passes a 100 micron sieve. Since the primary route of exposure is inhalation, the Department is considering only regulating elemental metals which have a particle size less than 10 microns. A sonic sifting method is available to measure particle size down to 5 microns, and a test of a metal-containing waste showed that the method can be used effectively on actual industrial wastes. The concentration of regulated metals would be calculated by the following formula:

regulated metal in waste (mg/kg) = metals conc in sub 10 micron fraction (mg/kg) * wt of sub 10-micron fraction (kg)/Total wt of sample (kg)

Wastes containing elemental mercury are hazardous regardless of concentration.

C. Modifications to Section 66261.24(a)(3): Acute Oral LD_{50}

It is important to retain the acute toxicity criterion to protect humans against the most immediate threats of illness or injury. Furthermore, acute toxicity is the only method within the California waste classification system of identifying toxic effects on humans of waste constituents which are not on the list of persistent and bioaccumulative compounds. DTSC recommends the adoption of two-tiered acute toxicity criteria to ensure that highly toxic substances are managed so as to avoid exposures at toxic levels while not over-regulating chemicals of moderate acute toxicity. DTSC has developed recommended acute toxicity thresholds for hazardous wastes and special wastes. The oral LD_{50} of a waste may be determined in a dosing study with laboratory animals or may be calculated from published LD_{50} values of the individual components of the waste using the following formula:

LD_{50} (waste) = 100/(weight % C_a/LD50 C_a + weight % C_b/LD50 C_b ...+ weight % C_n/LD_{50} C_n)

APPENDIX D: DTSC REPORT

where C_n represents the nth acutely toxic chemical in the waste. Hazardous wastes would include wastes with a calculated or experimentally determined oral LD_{50} less than 30 mg/kg. Non-hazardous wastes would include those with an oral LD_{50} exceeding 500 mg/kg. Special wastes would include wastes with oral LD_{50}s between 30 and 500 mg/kg. Derivation of these values is in Appendix 4.

D. Modifications to Section 66261.24(a)(4): Acute Dermal LD_{50}

Hazardous wastes would include wastes with a calculated or experimentally determined dermal LD_{50} less than 5500 mg/kg. Non-hazardous wastes would include those with a dermal LD_{50} exceeding 7400 mg/kg. Special wastes would include wastes with dermal LD_{50}s between 5500 and 7400 mg/kg. Derivation of these values is on page 45 in Appendix D. The formula for calculating the dermal LD_{50} of a waste is:

$$LD_{50} \text{ (waste)} = 100/(\text{weight \%} \ C_a/LD_{50} \ C_a + \text{weight \%} \ C_b/LD_{50} \ C_b \ ... + \text{weight \%} \ C_n/LD_{50} \ C_n)$$

where C_n represents the nth acutely toxic chemical in the waste.

E. Modifications to Section 66261.24(a)(5): Acute Inhalation LC_{50}

Under the proposed classification system, wastes would need to be evaluated for risk of acute toxic effects from inhalation exposures to both volatiles and particulates. The highest category determines the classification of the waste.

<u>Volatiles:</u> In order to account for both a chemical's acute inhalation toxicity and its tendency to vaporize, classification of a waste containing volatile constituents would be based on the ratio of each constituent chemical's vapor pressure (in ppm @ 25° C) to its inhalation LC_{50} (in ppm). If this ratio exceeds 0.1, the waste containing the chemical would be a special waste. If this ratio exceeds 1, the waste containing the chemical would be a hazardous waste. These ratios must be summed for wastes with multiple volatile chemicals, i.e. $\Sigma(VP/LC_{50}) > 0.1$ yields a special waste classification and $\Sigma(VP/LC_{50}) > 1$ yields a hazardous classification. In order to calculate this ratio, vapor pressure in mm Hg is converted to vapor pressure in atmospheres by dividing by 760. This, in turn is converted to ppm by multiplying by 1 million.

<u>Particulates:</u> Classification of a waste based on its particulate constituents would be based on the respirable fraction of the waste (the fraction with a particle size less than 10 microns) times the sum of the ratios of each chemical's concentration (in mg/kg) in the respirable fraction divided by its inhalation LC_{50} (in mg/m³). This C/LC_{50} ratio accounts for the tendency of the chemical to be suspended in the air and for its acute toxicity by inhalation. The table below is used to classify the waste:

Vapor Pressure/LC_{50} ratio sum	Classification	Concentration/LC_{50} sum
VP/LC < 0.1	Non-hazardous waste	$C/LC_{50} < 10^5$
0.1 < VP/LC < 1	Special Waste	not applicable
VP/LC > 1	Hazardous Waste	$C/LC_{50} > 10^5$

Derivation of these values is in Appendix D, page 45.

2/27/98

F. Modifications to Section 66261.24(a)(6): Aquatic Toxicity

The Department recommends retaining the fish bioassay and the current exit threshold. The fish bioassay is important because it is the only method within the California waste classification system of identifying ecotoxic properties of waste constituents which are not on the list of persistant and bioaccumulative compounds and it is the only method for evaluating the integrated toxic effects of multiple constituents of a waste. Wastes with aquatic $LC_{50}s$ below 500 mg/l, but above 30 mg/l would be classified as special wastes. With concurrence from the Reqional Water Quality Control Board, special wastes could be disposed of in lined municipal solid waste landfills meeting current RCRA Subtitle D requirements for new landfills. Wastes with aquatic $LC_{50}s$ below 30 mg/l would be classified as hazardous wastes and would require disposal in Class 1 landfills.

G. Modifications to Section 66261.24(a)(7): Listed Carcinogens

The Department recommends that vinyl chloride be removed from the carcinogenicity section, since it is being added to the TTLC and SERT lists.

H. Modifications to Section 66261.24(a)(8): New Threats

The Department recommends that this criterion be retained. As industrial processes and consumer products evolve, there will be new chemicals in commerce and new opportunities for exposures to humans or non-human biota. Without the ability to identify as hazardous those wastes which have been shown to be toxic by testing or experience, the time required to promulgate new regulation may prevent the Department from acting quickly to remedy a situation arising from a new chemical, a new type of exposure, or new knowledge.

I. Classification of wastes under the proposed regulatory structure

Figure 1 on the following page is a diagram of the proposed non-RCRA classification structure, excluding sections 66261.24(a)(7) and (a)(8). **Note:** Nothing in this document shall be construed to amend or repeal current regulations. All current regulations remain in effect unless and until DTSC has completed rulemaking and a new requirement is formally adopted.

Reclassifications and Variances

The TTLCs, SERTs, and acute toxicity thresholds are rigid criteria applied to all wastes regardless of their location or physical characteristics. Not all wastes containing a concentration of chemical exceeding one or more of these criteria are likely to adversely affect humal health or the environment. The physical characteristics of the waste and its management greatly influence the potential hazard. The Department is called on to reclassify wastes or grant variances based on physical characteristics and/or management practices. Defining a process for computing a revised set of TTLCs and SERTs based on multimedia risk assessment would establish a logical basis for reclassifying wastes using applicant-supplied waste-specific or site-specific parameters in the model. DTSC proposes to use the paradigms described herein as a conceptual basis for analyzing mitigating properties or circumstances in review of applications for variances and reclassifications.

APPENDIX D: DTSC REPORT

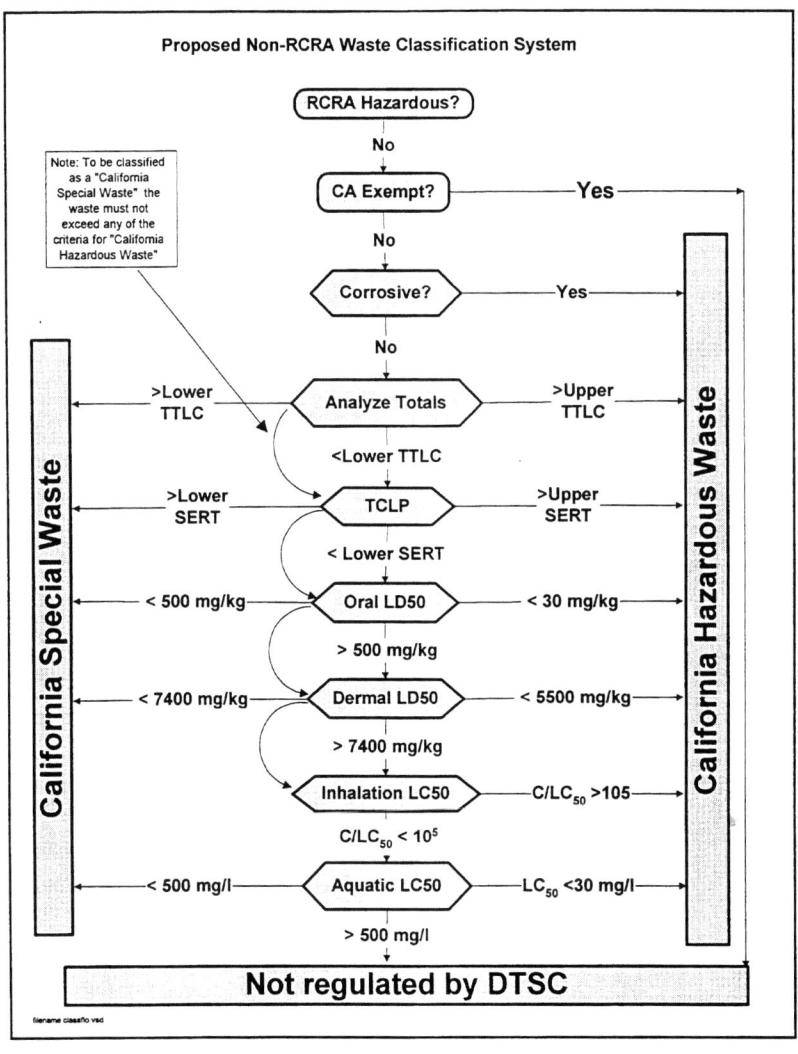

Appendix 1: Regulated chemicals

ORGANIC CHEMICALS

Aldrin is a chlorinated insecticide that is no longer produced in or imported into the United States. It is closely related to dieldrin chemically and is readily converted to dieldrin in the environment. It is dieldrin which bioaccumulates in the environment. It is relatively insoluble in water and persists in soil. Aldrin is well absorbed through the skin and is acutely toxic to mammals, fish, and birds. The Cal/EPA oral and inhalation cancer potency factor is 17 (mg/kg-day)$^{-1}$..Aldrin also causes non-cancer liver effects at very low doses equivalent to the 10^{-5} cancer risk level.

Chlordane is a chlorinated insecticide that is no longer used in the United States because of its adverse health effects and persistence in the environment. However, it is still manufactured for export. It has been found in water, soil, and air as it was widely used as a termiticide. Chlordane is absorbed through the skin and mucous membranes and is bioaccumulated in human body fat. Chlordane is also bioaccumulated in fish, birds, and mammals. Chlordane is highly toxic to aquatic organisms. The Cal/EPA oral and inhalation cancer potency factor is 1.2 (mg/kg-day)$^{-1}$, based on liver tumors in laboratory animals. However, chlordane causes non-cancer liver effects at very low doses and the TTLCs are determined by these effects because they occur at lower doses than the liver tumors. It is also important to realize that non-cancer effects are based on a maximum annual concentrations rather than a time-weighed average over the exposure duration. The maximum dose occurs via the breast milk pathway during the first year of life.

DDT (dichloro diphenyl trichloroethane) is an insecticide that has been banned for agricultural uses in the United States since 1972 because of its environmental persistence, accumulation in the food chain, interference with avian reproduction, and carcinogenicity in laboratory animals. It is insoluble in water and binds tightly to soil particles. DDT is moderately acutely toxic to mammals, fish, and birds. The Cal/EPA oral and inhalation cancer potency factor is 0.34 (mg/kg-day)$^{-1}$. DDT is transformed in the environment to DDE and DDD. The toxicity and movement of these three compounds is similar, though not identical.

2,4-D (2,4-Dichlorophenoxyacetic Acid) is a chlorinated organic acid herbicide that does not persist in soil, but can move readily through soil to groundwater. It is nontoxic to fish. The U.S. EPA reference dose is 0.01 mg/kg-day, based on blood, liver, and kidney effects in laboratory animals. Like pentachlorophenol and silvex, 2,4-D represents a challenge for CalTOX because it is an ionizable compound. This means it can exist in either if two states with very different chemical properties. At low pH values it behaves like a any of the other non-ionizable chemicals with some vapor pressure and low solubility in water. At high pHs, it is ionized and therefore, has a very low vapor pressure (unmeasurable, like metals), but is much more water soluble. We have used the chemical parameters which characterize the unionized form of 2,4-D. This was done because the objective of the TTLCs is to characterize the non-groundwater pathways and the unionized form moves more rapidly in those pathways. The SERTS are designed to address the groundwater pathways. Since 2,4-D is a herbicide, the backyard garden pathway has been eliminated as a potential route of exposure for both upper and lower TTLC calculations.

Dieldrin is a chlorinated insecticide that is no longer produced in or imported into the United States. It is easily converted from aldrin in the environment and is more resistant to biodegradation than aldrin. Dieldrin is insoluble in water, and binds strongly to soil particles, and biomagnifies through the food chain. It is acutely toxic to mammals and especially to fish. The Cal/EPA oral and inhalation cancer potency factor is 16 (mg/kg-day)$^{-1}$.

APPENDIX D: DTSC REPORT

Endrin is a chlorinated insecticide that is no longer produced or used in the United States because of its toxicity to birds. It persists in the environment in sediments and soil and bioaccumulates in fatty tissue of animals. It is toxic to aquatic organisms. The U.S. EPA reference dose is 0.0003 mg/kg-day, based on liver effects.

Heptachlor is a chlorinated insecticide used as a termiticide in the United States. Its use is limited because of concern over its carcinogenic potential. It is moderately persistent in soil, can bioconcentrate in and is toxic to aquatic organisms. The Cal/EPA oral and inhalation cancer potency factor is 5.7 (mg/kg-day)$^{-1}$, based on liver tumors in laboratory animals.

Kepone (chlordecone) is a chlorinated insecticide, related to mirex, that has not been allowed for use in the United States since 1978 and is no longer produced in the United States. It is relatively insoluble in water, binds strongly to soil particles, and has a very high potential to bioaccumulate in fish and other aquatic organisms. Kepone is moderately acutely toxic to mammals. The Cal/EPA oral and inhalation cancer potency factor is 16 (mg/kg-day)$^{-1}$.

Lead, organic: The gasoline additive, tetraethyl lead, was used as the prototype for organic lead compounds. The reference dose for tetraethyl lead is 10^{-7} mg/kg.

Lindane (gamma HCH) is a chlorinated insecticide which is commonly used to treat head and body lice and scabies, although it is no longer manufactured in the United States. Lindane is not persistent in soil but can remain in the air for some time. It can be absorbed through the skin, lungs, and gastrointestinal tract. In the body it can break down to pentachlorophenol. Lindane is toxic to fish and birds. The Cal/EPA considers lindane to be a carcinogen with an oral and inhalation cancer potency factor of 1.1 (mg/kg-day)$^{-1}$.

Methoxychlor is a chlorinated insecticide manufactured in the United States and widely used as a replacement for DDT because of its low toxicity in animals and humans. It is somewhat persistent in the environment. The targets of its toxic effects in humans are the neurological and reproductive systems. It is toxic to fish. The U.S. EPA reference dose is 0.005 mg/kg-day, based on maternal toxicity.

Mirex is a chlorinated insecticide that is chemically related to kepone and is no longer used or made in the United States. It is practically insoluble in water, binds tightly to soil particles, and has been shown to bioaccumulate in fish, other aquatic organisms, and plants. It is moderately acutely toxic to mammals. The Cal/EPA oral and inhalation cancer potency factor is 18 (mg/kg-day)$^{-1}$.

Pentachlorophenol (PCP) is a chlorinated molluscicide, fungicide, and wood preservative and is manufactured and used in the United States. Since it can exist in the environment in either the nonionized or ionized state, pentachlorophenol can exhibit a range of behavior in environmental media. However, it is not as persistent as some other chlorinated chemicals. It bioaccumulates in fish and other aquatic organisms. The proposed TTLC is based on the Cal/EPA oral and inhalation cancer potency factor of 0.018 (mg/kg-day)$^{-1}$. The RCRA TC threshold is based on non-carcinogenic effects. Like 2,4-D and silvex, PCP represents a challenge for CalTOX because it is an ionizable compound. This means it can exist in either if two states with very different chemical properties. At low pH values it behaves like a any of the other non-ionizable chemicals with some vapor pressure and low solubility in water. At high pHs, it is ionized and therefore, has a very low vapor pressure (unmeasurable, like metals), but is much more water soluble. We have used the chemical parameters which characterize the unionized form of PCP. This was done because the objective of the TTLCs is to characterize the non-groundwater pathways and the unionized form moves more rapidly in those pathways. The SERTS are designed to address the groundwater pathways.

PCBs (polychlorinated biphenyls) are a class of persistent chemicals formerly used as coolants, lubricants, and dielectric fluids in electrical equipment. They have not been manufactured in the United States since 1977 because of their toxicity, carcinogenicity, and environmental persistence. PCBs are highly insoluble in water, bind strongly to soil, bioaccumulate in aquatic organisms and have been shown to biomagnify up the food chain. They are carcinogenic and have reproductive and endocrine effects, and are toxic to aquatic and terrestrial organisms. PCBs are a mixture of chemical isomers with a range of physical, chemical and toxicological properties. A number of different mixtures have been identified for toxicological/regulatory purposes. The chemical and physical properties used were those of PCB-1254. The Cal/EPA oral and inhalation cancer potency factor is 7.7 (mg/kg-day)$^{-1}$, but the TTLCs are determined by non-cancer liver effects, which occur at lower doses than the cancer effects. It is also important to realize that non-cancer effects are based on a maximum annual concentrations rather than a time-weighed average over the exposure duration. For chlordane and PCBs, the maximum concentration occurs via the breast milk pathway during the first year of life. PCBs are regulated by the federal government under the Toxic Substances Control Act, with a regulatory threshold of 50 ppm.

Dioxins are a class of chemicals that have never been intentionally manufactured but are generated as byproducts from combustion and chemical processes. They affect the reproductive and endocrine systems. Dioxins are insoluble in water, bind strongly to soil particles, and are highly resistant to degradation. They are accumulated by aquatic organisms, providing entry into the aquatic food chain. Dioxins in air deposit onto plants, providing entry into the terrestrial food chain. The Cal/EPA oral and inhalation cancer potency factor is 130,000 (mg/kg-day)$^{-1}$. 2,3,7,8 TCDD, the most potent isomer, is regulated under Title 22.

Toxaphene is a chlorinated insecticide which is no longer used in the United States because it is toxic to humans, fish, and other animals. Toxaphene persists in soil and sediments and is resistant to breakdown in the environment. It is easily absorbed through the skin and lungs. The Cal/EPA oral and inhalation cancer potency factor is 1.2 (mg/kg-day)$^{-1}$, based on liver and thyroid tumors in laboratory animals.

Trichloroethylene (TCE) is a chlorinated solvent mostly used to degrease metal parts. It can move readily through soil to groundwater and can evaporate into the air. It is minimally toxic to aquatic organisms. The Cal/EPA considers TCE a potential human carcinogen with an oral cancer potency factor of 0.015 (mg/kg-day)$^{-1}$ and an inhalation cancer potency factor of 0.01 (mg/kg-day)$^{-1}$.

2,4,5-trichlorophenoxy propionic acid (silvex) is an herbicide which is no longer used in the United States because of health effect concerns, particularly birth defects. This chemical does not appear to be of ecological concern. The U.S. EPA reference dose is 0.008 mg/kg-day, based on liver changes in laboratory animals.

Like pentachlorophenol and 2,4-D, silvex represents a challenge for CalTOX because it is an ionizable compound. This means it can exist in either if two states with very different chemical properties. At low pH values it behaves like a any of the other non-ionizable chemicals with some vapor pressure and low solubility in water. At high pHs, it is ionized and therefore, has a very low vapor pressure (unmeasurable, like metals), but is much more water soluble. We have used the chemical parameters which characterize the unionized form of silvex. This was done because the objective of the TTLCs is to characterize the non-groundwater pathways and the

2/27/98

APPENDIX D: DTSC REPORT

unionized form moves more rapidly in those pathways. The SERTS are designed to address the groundwater pathways. Finally, silvex is a herbicide, therefore, the backyard garden pathway has been eliminated as a potential route of exposure for both upper and lower TTLC calculations.

Vinyl chloride is a volatile chlorinated chemical used to make many plastic products and is most readily released to air. It is soluble in water and thus can leach through soil to groundwater. It is considered toxic to aquatic organisms. Vinyl chloride is one of the few chemicals to have been shown to cause cancer in humans. The Cal/EPA oral and inhalation cancer potency factor is 0.27 (mg/kg-day)-1, based on a rare form of liver tumors in laboratory animals.

INORGANIC CHEMICALS

Antimony is a metal used in alloys; the oxide is used as a fire retardant. Elevated atmospheric levels can irritate eyes, lungs, and skin. Antimony is a gastrointestinal irritant. Systematically it can cause anemia, electrocardiographic anomalies and muscle or joint pain. Antimony causes reproductive and developmental effects in non-human species. The U.S. EPA reference dose is 0.0004 mg/kg-day.

Arsenic is an element used in many industrial processes. It has acute and chronic toxic effects and is a human carcinogen. The U.S.EPA oral cancer potency factor is 1.5 (mg/kg-day)$^{-1}$.

Barium is an element with various medical and industrial uses. It has low toxicity to aquatic organisms. The U.S. EPA reference dose is 0.07 mg/kg-day, based on increased blood pressure in humans.

Beryllium is a metal with a variety of uses in metallurgy and consumer and industrial products. It causes acute and chronic berylliosis when inhaled, causes cancer in laboratory animals, and is toxic to aquatic organisms. The Cal/EPA oral and inhalation cancer potency factor is 7 (mg/kg-day)$^{-1}$.

Cadmium is a metal used in batteries, metal plating, pigments, photography and lithography. It is released into the environment mostly through mining and refining practices and from incineration of coal and wastes. Like most other elements, cadmium is not well absorbed through the skin. Cadmium can bioaccumulate and is toxic to aquatic and terrestrial organisms. The Cal/EPA inhalation cancer potency factor is 15 (mg/kg-day)$^{-1}$, based on respiratory tumors.

Chromium is an element that is an essential nutrient for humans. It is also produced by industrial processes, particularly plating, dyes and pigments, leather tanning and wood preserving. Chromium most often enters the body through ingestion and inhalation. The U.S. EPA reference dose for trivalent chromium is 1 mg/kg-day, based on no effects seen in a rat feeding study.

Hexavalent chromium is produced by industrial processes, particularly plating, dyes and pigments, leather tanning and wood preserving. Chronic inhalation of hexavalent chromium can irritate the nasal passages and lead to sensitization or lung cancer. Ingested hexavalent chromium can irritate the gastrointestinal tract and lead to toxic manifestations in the liver and kidney. Hexavalent chromium is toxic to aquatic and terrestrial organisms. The Cal/EPA oral cancer potency factor is 0.42 (mg/kg-day)$^{-1}$. The Cal/EPA inhalation cancer potency factor is 510 (mg/kg-day)$^{-1}$.

Cobalt is a metal used in alloys, in porcelain and in pigments. It is an integral part of Vitamin B$_{12}$, an essential nutrient. It is present in the environment as a result of natural processes and of burning fossil fuels. Inhalation of excessive amounts of cobalt can cause asthma and

2/27/98

pneumonia. Ingestion of excessive amounts of cobalt has resulted in chronic cardiomyopathy. The U.S. EPA reference dose (currently under review) is 0.06 mg/kg-day.

Copper is a soft reddish metal used for many industrial and consumer products by itself or in alloys or in the form of copper compounds. It is an essential nutrient. Long-term exposure to elevated levels of copper can irritate mucous membranes and cause headaches or dizziness. Copper's greatest threat in the environment is its toxicity to aquatic organisms. Although many copper salts are soluble, in natural systems copper is usually bound to soil or sediment. The U.S. EPA reference dose is 0.037 mg/kg-day. The U.S. EPA Ambient Water Quality Criterion is 12 ug/l.

Fluoride is a halogen whose major use is addition to drinking water at concentrations around 1 mg/l as an aid in the prevention of dental caries. Excessive dosages can cause dental and skeletal fluorosis, a mottling of teeth and bones. This does not become a significant clinical problem until the dosage reaches 20-40 mg/day. The U.S. EPA reference dose is 0.06 mg/kg-day.

Lead is a metal which has been used for numerous industrial purposes and is present in the environment mostly from the past combustion of leaded gasoline. Lead is readily absorbed across the gastrointestinal tract into the blood stream and then many tissues in the body. Chronic exposure by humans has an adverse effect on many organs, most importantly, the central nervous system, even at very low concentrations. The Center for Disease Control has stated that children's blood lead level should be below 10 ug lead/dl.

Mercury is a metal used as a fungicide, in electrical apparatus, and other industrial purposes. Unique among metals, mercury can evaporate at room temperature; in this way, mercury easily enters the atmosphere, where it can be inhaled or redeposit onto soil and surface water. Mercury is toxic to aquatic and terrestrial organisms. The U.S. EPA reference concentration is 0.0003 mg/m3, based on central nervous system effects after inhalation.

Molybdenum is a metal used in making alloys, particularly where high temperature resistance is needed. It is an essential nutrient. Excessive dosages can cause reversible anemia, diarrhea, and poor growth. Higher or prolonged exposures can lead to joint deformities and degenerative changes in the liver and kidney. Copper antagonizes the absorption of molybdenum. Molybdenum is toxic to aquatic organisms. The U.S. EPA reference dose is 0.005 mg/kg-day.

Nickel is an important industrial metal used in alloys, plating, and batteries. Nickel exposures are most common in workers in these industries. Acute exposures to nickel carbonyl result in respiratory and generalized symptoms resembling viral pneumonia. Dermal contact can lead to dermatitis and hypersensitivity. Nickel subsulfide, nickel refinery dust, and probably nickel carbonyl are carcinogenic when inhaled. Chronic exposure can lead to impaired immune function. Nickel is toxic to aquatic organisms. The U.S. EPA reference dose is 0.02 mg/kg-day. The acute toxicity endpoint was allergic dermatitis in humans, reported to occur at a LOEL of 0.01 to 0.1 mg/kg.

Selenium is an element used in electronic equipment and other industrial products. It is released to the environment via natural and manufacturing processes. Humans are mostly commonly exposed to selenium through the ingestion of selenium-containing foods. Selenium is moderately toxic to aquatic organisms and very toxic to certain birds. The U.S. EPA reference dose is 0.005 mg/kg-day, based on selenosis in humans.

Silver metal and its compounds are used in photographic materials, electrical products, alloys and jewelry. Silver does not break down in the environment and is present at low concentrations in water and soil. Most of the silver released to the environment comes from photographic materials or from mining operations. Free silver ions are very toxic to aquatic organisms, but are not persistent in the environment. The U.S. EPA reference dose is 0.005 mg/kg-day, based on argyria, a discoloration of the skin in humans who have absorbed a cumulative dose of at least 1 gm silver (equivalent to at least 25 gm ingested). Daily dosages associated with toxic effects in rats are much higher.

Thallium is a metal used in the electronics industry, in switches, in specialized glass, and in some medical devices. Its use as a rat poison has been banned. It is readily taken up by plants and is relatively well absorbed from the gastrointestinal tract. Toxic effects may be found in the nervous system, and the heart, liver, and kidneys. The U.S. EPA reference dose is 0.00007 mg/kg-day.

Vanadium is a metal that is used in steel-making, plastics, ceramics, and rubber. It is released to the atmosphere when fossil fuels are burned. Principal health effects involve the respiratory system, producing such symptoms as cough, chest pain, and sore throat. Vanadium is toxic to aquatic organisms. The U.S. EPA reference dose is 0.003 mg/kg-day.

Zinc is a metal that has many industrial uses by itself or in alloys. It is released to the environment from mining and industrial operations, fossil fuel combustion, and incineration of zinc-containing wastes. It is an essential nutrient. Zinc may cause digestive upsets associated (often from overdosing with zinc supplements), or impairment of pulmonary function associated with inhalation of zinc dust or fumes. The U.S. EPA reference dose is 0.3 mg/kg-day. Zinc is toxic to fish at relatively low concentrations. Aqueous solubility depends on pH, salinity, and presence of ligands and complexing agents. The U.S. EPA Ambient Water Quality Criterion is 110 ug/l.

Appendix 2: Derivation of proposed SERTs

Lower (exit-level) SERTs

The proposed lower Soluble or Extractible Regulatory Thresholds (SERTs) are based on the lowest of:

(1) California Maximum Concentration Limits (MCLs), or

(2) U.S. EPA Ambient Water Quality Criteria for the protection of aquatic life at 100 ppm total hardness or a pH of 6.5 (as applicable), or

(3) human-health-based levels calculated by DTSC. These values consider only exposure to humans from drinking ground water.

To calculate health-based levels for carcinogenic constituents, cancer potency factors developed by the Office of Environmental Health Hazard Assessment within the California Environmental Protection Agency (if available, otherwise U.S. EPA cancer potency factors) were used to calculate the concentration corresponding to a risk of 10^{-5}. This risk level was chosen because it is the no-significant-risk level under the Safe Drinking Water and Birth Defects Prevention Act. The balance of the health-based values are based on U.S. EPA reference doses or on data used to develop the U.S. EPA region 9 Preliminary Remediation Goals (PRGs) for non-carcinogenic effects. Carcinogenic effects of lead were considered but were not limiting. The general equation for the health-based levels for carcinogens can be written as:

$$HBL = Risk / (WI \times CPF \times ED/AT)$$

The general equation for the health-based levels for non-carcinogens can be written as:

$$HBL = RfD/WI_c$$

The parameters and their distributions are as follows:

Parameter	mean	st. dev.
AT (lifespan, years)	70*	7*
ED (exposure duration, yrs)	14*	16*
WI (water ingestion, combined, l/kg/day)	0.022*	0.004*
WI_c (water ingestion, child, l/kg/day)	0.029*	0.01*
Cancer Potency Factors (kg-day/mg)	Cal/EPA (chemical-specific)	
Reference dose ($mg \cdot kg^{-1} \cdot day^{-1}$)	U.S. EPA RfD or region 9 PRG	

* Values from CalTOX

All health-based values are 10th percentile estimates of the concentrations that would correspond to the stated level of risk or hazard. This means that for an individual picked at random from an exposed population there is theoretically a 10% chance that the true risk is higher than 10^{-5} for carcinogens or that the true hazard index is larger than 1 for non-carcinogens. Conversely, there is a 90% chance that the true risk is lower than 10^{-5} or that the

APPENDIX D: DTSC REPORT 189

true hazard index is less than 1. The 10th percentile was chosen to balance protectiveness and practical considerations.

A limitation of this probabilistic approach is that it is only as accurate as the distributions that are put into the model. The quantity and quality of data from which to construct parameter distributions is highly variable. Unfortunately, there are no widely accepted distributions for the toxicity parameters, so conservative estimates are used: 1) Cancer potency slopes are statistical 95th percentile estimates of the slope of the line describing the relationship between carcinogen dose and probability of cancer in the test species; 2) the most sensitive species and sex are used (unless there are satisfactory human data); 3) humans are usually assumed to be 6 times as sensitive to carcinogens as rats and 14 times as sensitive as mice (when rodent studies are relied upon; and 5) it is assumed that at low dosages the relationship between the exposure to the carcinogen and the probability of neoplasia is linear with no threshold. For non-carcinogens, uncertainty factors are usually incorporated to compensate for possible differences between species and to protect the most sensitive individual. For these reasons, there is much less than a 10% chance that the true risk is as high as 10^{-5} or that the true hazard index is larger than 1, but this extra conservatism can only be considered qualitatively.

The calculated values also include the implicit assumption that the waste containing the constituent in question is undiluted in the landfill. This is a conservative assumption because, in reality, most waste would be mixed with other wastes which do not contain the constituent being evaluated (unless the waste is disposed of in a monofill, in which case the assumption that the waste is co-disposed with decomposing organic matter would not apply), and leachate which came in contact with the waste would be diluted by leachate which did not contain the constituent in question. Thus, it is assumed that the concentration of the constituent in the TCLP extract is a good representation of the concentration of the constituent in landfill leachate after mixing with leachate from the entire landfill cell.

The calculated values consider ingestion of groundwater as the only human exposure pathway. Bioaccumulation (increasing body burden over time) is taken into account because the toxicity benchmarks are usually based on studies involving lifetime exposure (and if not, uncertainty factors are incorporated to account for less-than-lifetime exposure).

To calculate the proposed SERT, the MCL, AWQC, or HBL (whichever was lowest) was multiplied by a dilution/ attenuation factor of 100. (This factor is the same as that used by U.S.EPA in developing the RCRA Toxicity Characteristic regulatory limits, and is intended to account for the dilution and attenuation that occur as leachates move through the unsaturated zone and mix with ground water. U.S. EPA is in the midst of a project to develop chemical-specific dilution/attenuation factors. DTSC will track the progress of this effort, and consider adopting chemical-specific dilution/attenuation factors when these are developed.) The resulting value was compared with the estimated quantitation limit (EQL) x 2, and the higher value (rounded to one significant figure) became the proposed SERT.

SERTs for antimony, arsenic, molybdenum, selenium, and vanadium are not intended to be applied to solid wastes because the extraction test (the Toxicity Characteristic Leaching Procedure or TCLP) does not reliably predict leaching of these elements from solid wastes.

SERTs that exceed 1/20 of the corresponding TTLC are not intended to be applied to solid wastes. This is because there is a 20-fold dilution inherent in the TCLP methodology and therefore a waste could not exceed a SERT without also exceeding the TTLC unless the SERT is less than 1/20 of the TTLC, even if the constituent is 100% soluble or extractable. All SERTs apply to liquid wastes unless the TTLC is lower.

The lower (exit-level) SERT for arsenic was calculated as follows: The health-based drinking water concentration limit, based on an oral carcinogenic potency of 1.5 $(mg/kg/day)^{-1}$ and a risk of 10^{-5}, was 0.007 mg/l. However, the ninetieth percentile arsenic concentration in California drinking water supplies in 1994 as measured by the Department of Health Services was 0.01 mg/l. DTSC's mandate to protect public health and the environment does not include regulation to below background concentrations, and therefore we propose to use 0.01 mg/l as the target arsenic concentration limit in ground water. With an assumed dilution/attenuation factor of 100, the limit in liquid waste would be 1 mg/l. In the following table, proposed SERTs are compared with other regulatory criteria.

APPENDIX D: DTSC REPORT

Basis for Lower SERTs and comparison with current regulatory thresholds (mg/l)

	Health-based level x 100	CA MCL x 100	AWQC x 100	EQL[h] x 2	proposed lower SERT	Current STLC	RCRA RL
Aldrin	0.006[a]	none	0.3	0.0007	0.006	0.14	none
Chlordane	0.002[a]	0.01	0.0004	0.0007	0.0007	0.25	0.03
DDT & metabolites	0.7[a]	none	0.0001	0.0005	0.0005	0.1	none
2,4-,D	30[b]	7	none	0.004	7	10	10
Dieldrin	0.01[a]	none	0.0002	0.0009	0.00099	0.8	none
Endrin	1[b]	0.2	0.0002	0.0008	0.0008	0.02	0.02
Heptachlor	0.04[b]	0.001	0.0004	0.0008	0.0008	0.47	0.008
Kepone	0.006[a]	none	none	0.04	0.04	2.1	none
Lindane	0.2[a]	0.02	0.008	0.0005	0.008	0.4	0.4
Methoxychlor	20[b]	4	0.003	0.002	0.003	10	10
Mirex	0.01[a]	none	0.0001	0.0003	0.0003	2.1	none
Pentachlorophenol	10[a]	0.1	0.4	0.002	0.1	1.7	100
PCB	0.03[a]	0.05	0.001	0.02	none[i]	5	none
2378 TCDD	2e-6[a]	3e-6	none	2e-9 to 2e-8	none[i]	0.001	none
Toxaphene	0.2[a]	0.3	0.00002	0.002	0.002	0.5	0.5
TCE	20[a]	0.5	none	0.01	0.5	204	0.5
2,4,5-trichlorophenoxy	30[b]	5	none	0.002	5	1	1
Vinyl chloride	3	0.05	none	0.01	0.05	5	5
Antimony	1[b]	0.6	3	1	1	15	none
Arsenic	0.07[a]	5	20	0.2	1e	5	5
Barium	200[b]	100	none	4	100	100	100
Beryllium	0.02[a]	0.4	0.5	0.1		0.75	none
Cadmium	2[b]	0.5	0.1	0.1	0.1	1	1
Chromium+6	0.2[a]	none	1	0.02	0.2	5	none
Chromium, total	3000[b]	5	20	0.2	5	560	5
Cobalt	200[c]	none	none	1	200	80	none
Copper	100[b]	none	1	0.5	1	25	none
Fluoride	200[b]	none	none	1	200	180	none
Mercury	1[b]	0.2	0.001	0.004	0.004	0.2	0.2
Molybdenum	10[b]	none	none	1	10	350	none
Nickel	70[b]	10	20	0.8	10	20	none
Lead	20[d]	1.5	0.3	0.06	0.3	5	5
Selenium	20[b]	5	0.5	0.1	0.5	1	1
Thallium	0.2[b]	0.2	40	0.2	0.2	7	none
Vanadium	20[b]	none	none	1	20	24	none
Zinc	1000[b]	none	10	0.4	10	250	none

7/2/98

Footnotes for preceeding and following tables

a. Based on California or U.S. EPA cancer potency factor with CalTOX exposure parameters.
b. Based on current U.S. EPA reference dose with CalTOX exposure parameters.
c. Based on U.S. EPA region 9 preliminary remediation goals
d. DTSC's lead risk assessment spreadsheet was used to determine the health-based level.
e. The proposed SERT for arsenic is based on CA background
f. Based on federal Ambient Water Quality Criteria @ 100 ppm hardness or a pH of 6.5 and longest available exposure
g. Based on estimated quantitation limit
h. No lower SERT proposed (see text)

Upper SERTs

The proposed upper SERTs were calculated by multiplying the lowest of the HBL, the MCL, or the AWQC by the DAF of 100 and by a liner protection factor (LPF). The liner protection factor reflects the assumption that wastes which are not classified as hazardous will require disposal in at least a single composite-lined landfill if they are land-disposed. The possible values of LPF (Tab 13) were entered as a custom distribution into the spreadsheet used to calculate SERTS. This distribution, along with the distributions for other variable parameters used in the SERT calculations, were propogated through the model to yield a distribution of possible values for each upper SERT. The resulting value is compared to the upper EQL x 2 and the higher value is the proposed upper SERT. As discussed earlier, the tenth percentile of this distribution is the proposed upper SERT.

APPENDIX D: DTSC REPORT

Basis for Upper SERTs and Comparison with Current Regulatory Thresholds (mg/l)

	HBL x 100	MCL x 100	AWQC x 100^f	EQL^n x 2	LPF	upper SERT	Current STLC	RCRA RL
Aldrin	0.006^a	none	0.3	0.007	63	0.4	0.14	none
Chlordane	0.002^a	0.01	0.0004	0.007	18	0.008	0.25	0.03
DDT & metabolites	0.7^a	none	0.0001	0.005	18	0.005	0.1	none
2,4-,D	30^b	7	none	0.04	18	100	10	10
Dieldrin	0.01^a	none	0.0002	0.009	18	0.009	0.8	none
Endrin	1^b	0.2	0.0002	0.008	18	0.008	0.02	0.02
Heptachlor	0.04^b	0.001	0.0004	0.008	18	0.008	0.47	0.008
Kepone	0.006^a	none	none	0.4	63	0.4	2.1	none
Lindane	0.2^a	0.02	0.008	0.005	18	0.1	0.4	0.4
Methoxychlor	20^b	4	0.003	0.02	18	0.05	10	10
Mirex	0.01^a	none	0.0001	0.003	18	0.003	2.1	none
Pentachlorophenol	10^a	0.1	0.4	0.02	18	2	1.7	100
PCB	0.03^a	0.05	0.001	0.2	18	5e	5	none
2378 TCDD	2e-6^a	3e-6	none	2e-8 to 2e-7	63	?	0.001	none
Toxaphene	0.2^a	0.3	0.00002	0.02	18	0.02	0.5	0.5
TCE	20^a	0.5	none	0.1	18	9	204	0.5
2,4,5-trichlorophenoxy	30^b	5	none	0.02	18	90	1	1
Vinyl chloride	3	0.05	none	0.1	18	0.9	5	5
Antimony	1^b	0.6	3	10	18	10	15	none
Arsenic	0.07^a	5	20	2	63	5	5	5
Barium	200^b	100	none	40	18	2000	100	100
Beryllium	0.02^a	0.4	0.5	1	63	1	0.75	none
Cadmium	2^b	0.5	0.1	1	18	2	1	1
Chromium+6	0.2^a	none	1	0.2	63	20	5	none
Chromium, total	3000^b	5	20	2	18	90	560	5
Cobalt	200^c	none	none	10	24	4000	80	none
Copper	100^b	none	1	5	18	20	25	none
Fluoride	200^b	none	none	10	24	4000	180	none
Mercury	1^b	0.2	0.001	0.04	18	0.04	0.2	0.2
Molybdenum	10^b	none	none	10	24	300	350	none
Nickel	70^b	10	20	8	18	200	20	none
Lead	20^d	1.5	0.3	0.6	18	5	5	5
Selenium	20^b	5	0.5	1	18	9	1	1
Thallium	0.2^b	0.2	40	2	19	4	7	none
Vanadium	20^b	none	none	10	24	500	24	none
Zinc	1000^b	none	10	4	18	200	250	none

7/2/98

Solubility, STLCs, and TTLCs

At present, there is no established laboratory procedure that satisfactorily estimates the leachability of antimony, arsenic, molybdenum, selenium, and vanadium from a variety of wastes. For these elements, DTSC evaluated the protectiveness of the TTLCs with respect to groundwater protection using the following approach: First, the fraction of the measured total concentration that is available for leaching was determined as follows. In a study conducted by DTSC, five composite wastes were analyzed for total concentration of the five elements and also reacted with leachates from seven landfills representing the spectrum of landfills in California. The extracted element concentration in those leachates were measured. For each leachate-waste combination a leachate-extractable concentration/total concentration ratio was computed. The highest ratio among all of the leachate-waste combinations (shown in the last column of the table below) represents the greatest extraction of the element by landfill leachate. Second, the maximum dissolved: total concentration ratio in the last column was multiplied by the proposed upper or entry TTLC. This computed maximum concentration in the leachate represents the leachate concentration that would result from the dissolution of the element present in the waste at the TTLC concentration with the most aggressive landfill leachate. The following table shows the proposed upper and lower TTLCs, the STLCs and computed maximum concentrations (in mg/l) in the leachates.

element	proposed upper TTLC (mg/kg)	STLC (mg/l)	computed maximum leachate conc.[1]	proposed lower TTLC (mg/kg)	STLC (mg/l)	computed maximum leachate conc.[1]	maximum dissolved: total ratio (mg/l÷mg/kg)
antimony	700	15	0.21	700	15	0.21	0.0003
arsenic	50	5	0.65	50	5	0.65	0.013
molybdenum	9,000	350	342	9,000	350	342	0.038
selenium	9,000	1	369	50	1	1.6	0.041
vanadium	10,000	24	180	2,000	24	36	0.018

1. The product of the TTLC and maximum dissolved:total ratio

Comparison of the adjacent shaded columns shows whether the predicted maximum extractable concentration exceeds the STLC. If it does, as in the case of the bolded values for selenium and vanadium, the TTLC would not be protective. In all other cases the computed maximum leachate concentration is less than the STLC. For selenium and vanadium, DTSC proposes to reduce the TTLCs to compensate for the inability of the TCLP to reliably predict the extraction of these elements from wastes. As shown below, TTLCs of 1000 for vanadium and 30 for selenium would result in maximum expected leachate concentrations which are less than STLCs.

element	proposed upper TTLC (mg/kg)	STLC (mg/l)	computed maximum leachate conc.	proposed lower TTLC (mg/kg)	STLC (mg/l)	computed maximum leachate conc.	maximum dissolved: total ratio (mg/l÷mg/kg)
selenium	30	1	1.2	30	1	1.2	0.041
vanadium	1,000	24	18	1,000	24	18	0.018

APPENDIX D: DTSC REPORT   195

Appendix 3: TTLCs

Introduction

What is the purpose of this appendix?

The concept paper includes a table of upper and lower Total Threshold Limit Concentration (TTLC) values for 38 chemicals. The concept paper describes how the upper and lower TTLC values are to be used in classifying a waste. This appendix describes the rationale for determining the values in the table.

What is the relationship between ground water and TTLCs?

The soluble or extractable fraction of the waste is that portion which may reach ground water. The TTLC tables and this appendix do not address the issues associated with the soluble fraction of waste (except in the case of chemicals for which the extraction test is found not to predict the leachable fraction). These are addressed by Soluble or Extractable Regulatory Thresholds (SERTs), described in Appendix 2.

Why are TTLCs developed for individual chemicals instead of waste streams?

The likelihood of adverse effects on health is proportional to the amount of a specific chemical people take into their bodies. Information is available in the scientific literature relating toxic effects in humans and animals to amounts of individual chemicals. The concentration in the waste of a specific chemical and published information about its toxicity are used to determine if a waste is hazardous. However, there is little scientific information on the toxicity of waste streams and additive or synergistic effects among individual chemicals. Testing of all California waste streams for toxic effects in laboratory animals would be neither cost-effective nor humane. Therefore, for purposes of comparison with TTLCs, wastes are analyzed for known toxic chemicals. The fish bioassay measures additive and/or synergistic effects.

Why were these 38 chemicals selected?

The objective of the Regulatory Structure Update project is to perform a critical review of current California hazardous waste regulations. Total Threshold Limit Concentrations (TTLCs) were determined to be important to retain because they have no counterpart in federal law. The Department believes that a multimedia approach to protection of public health and the environment is necessary. This effort is intended to improve the scientific basis of the TTLCs for the 38 chemicals which currently have TTLCs. Vinyl chloride has been added to the list because it is a chemical of commercial relevance which appears on the list of carcinogens in the current regulations.

What is the purpose of two tiers of TTLCs?

TTLCs (along with the other criteria) are intended to facilitate classification of wastes according to the hazard they pose. Such classification will allow us to determine which wastes should be regulated in order to protect people and ecosystems from exposures to toxic chemicals during the handling and disposal of wastes. TTLCs should protect workers and residents from chronic health effects (effects over long periods of time). The two-tier TTLCs would be used to classify wastes into three groups: 1) fully regulated hazardous wastes, 2) special wastes, with reduced regulatory or 3) no regulatory requirements imposed by DTSC. Wastes with concentrations above the upper TTLC value would be in group 1, wastes with concentrations lower than the lower TTLC would be in group 3 and wastes with any constituent concentration levels between the upper and lower TTLCs would be in group 2, provided that they were not classified in group 1 by any other characteristic.

How are the upper and lower (exit-level) TTLCs selected?

The proposed TTLC concentration for each chemical was selected from among several different concentrations in a two-step process that was nearly identical for both upper and lower TTLCs. First, the lower of two risk-based concentrations intended to protect human health and/or the environment was selected for each chemical. Second, the risk-based concentration was compared with two practical limitations: the limit of quantitation and the ambient concentration in the environment. The highest of the risk-based, ambient or quantitation limit concentration was selected for each chemical. The following diagram shows the decision process for selection of the TTLCs.

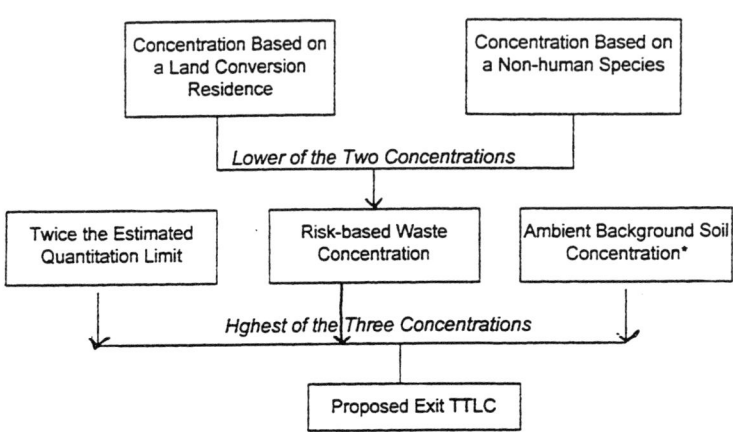

* Comparisons conducted for dioxin and inorganic chemicals

Figure 1: Decision Tree for Determining Exit TTLCs for Each Chemical

Figure 1 shows that for each chemical a decision was made to select a concentration that would protect both residents and non-human species living on or near land to which waste containing the TTLC chemical had been added. The scenario for residents is called the Land Conversion Scenario because the land is converted from uninhabited land to inhabited residential lots after waste ceases to be plowed into the soil. Once the lower of the two risk-based values was selected, this value was compared to both an estimated quantitation limit and, if available, an ambient concentration in soil. A policy decision was made to use the highest of these three concentrations. Thus, regulatory thresholds which are limited by quantitation limits or ambient concentrations could pose a theoretical risk exceeding the target risk to human health or the environment. The reasons for these decisions are as follows:

- Twice the Estimated Quantitation Limit was selected if it exceeded the risk-based value because regulated industries must be able to measure the TTLC concentration in order to determine if one is complying with the law.

- It is not reasonable or practical to define large amounts of native California soil as hazardous. Thus, if the risk-based concentration of any element was less than the maximum concentration measured in 50 samples of native California soils, a value based on the native soil ambient concentration was selected.

More detailed descriptions of the detection limit and ambient soil concentrations can be found under the

APPENDIX D: DTSC REPORT

appropriate subheadings below, and under tabs 5a and 5b.

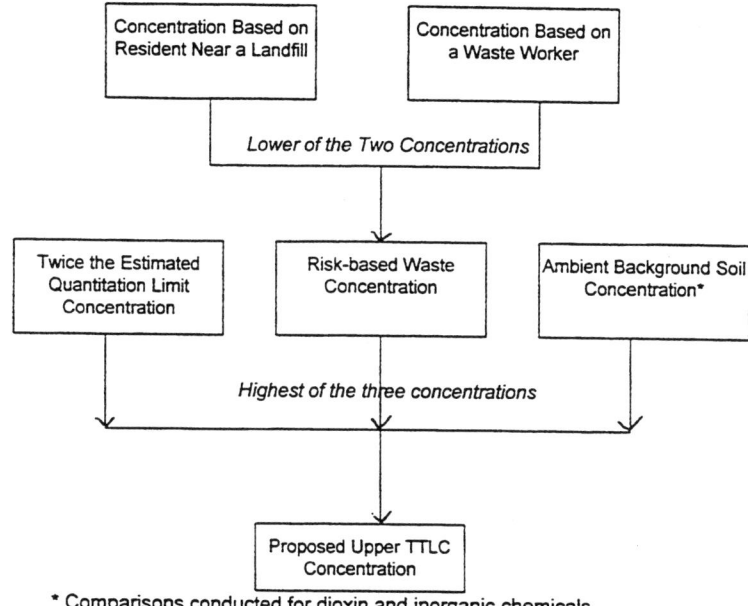

* Comparisons conducted for dioxin and inorganic chemicals

Figure 2: **Decision Tree for Determining Upper TTLCs for Each Chemical**

Figure 2 shows that for each chemical a decision was made to select a concentration that would protect both residents living near a facility in which waste containing the chemical is handled, and workers handling that waste. The computation of the concentrations based on the residential Landfill Scenario and the Waste Worker are described below under subheadings with those titles. Once the lower of the two risk-based values was selected, this value was compared to both an estimated quantitation limit and, if available, an ambient concentration in soil.. A policy decision was made to use the highest of these three concentrations. Thus, regulatory thresholds which are limited by quantitation limits or ambient concentrations could pose a theoretical risk exceeding the target risk to human health or the environment. The reasons for these decisions are as follows:

- Twice the Estimated Quantitation Limit was selected over a risk-based value because regulated industries must be able to measure the TTLC concentration, otherwise, it is not possible to determine if one is complying with the law.

- Worldwide ambient concentrations of dioxins and dibenzofurans exceed calculated health-based concentrations. Therefore, a value based on U.S. and U.K. ambient concentrations was selected as the basis for the proposed TTLC.

More detailed descriptions of the detection limit and ambient soil concentrations can be found under the appropriate subheadings below, and under tabs 5a and 5b.

First Stage: Computation of Risk-Based TTLCs

How is quantitative risk assessment used to assess effects on human health and the environment?

Risk assessment is the process of establishing a relationship between a chemical source (such as waste in a municipal landfill) and an adverse effect (such as risk of human cancer). A quantitative risk assessment establishes a mathematical relationship between a concentration of a chemical at a source, such as the soil in a residential lot or waste in a landfill, and the risk of an adverse health effect in people living on or near the source. Risk assessment comprises two parts:

1) Toxicity Assessment
Toxicity assessment pertains to the relationship between the risk of an adverse health effect and the dose or amount of chemical taken into the body. Knowledge about this relationship comes from toxicological studies (laboratory experiments with animals) and epidemiological studies (surveys of the health condition of humans). Different chemicals may induce different adverse health effects, but there is a relationship between the dose of the chemical and the likelihood or risk of the undesired effect. Toxicity assessment is chemical-dependent and independent of the exposure scenario.

Toxicity assessment is performed by the California Office of Environmental Health Hazard Assessment (OEHHA) and the U.S. EPA Office of Research and Development (USEPA-ORD). OEHHA and the USEPA have established a mathematical relationship between dose and the likelihood of getting cancer. Such a relationship between dose and risk of cancer is called a cancer potency factor. If an OEHHA potency factor exists for a specific chemical, this value is used in relating risk to dose. If no OEHHA potency factor exists, the USEPA potency factor is used. The USEPA has established reference doses (RfDs) that relate risk of non-cancer effects to a given dose. DTSC uses USEPA RfDs to establish maximum allowable daily doses for chemicals that cause non-cancer effects.

2) Exposure Assessment
Quantitative exposure assessment is the relationship between the source (such as a chemical concentration in a municipal landfill) and the dose a person may receive. Exposure assessment is based on an exposure scenario, such as a resident living near a municipal landfill. Sometimes these relationships are described by simple algebraic equations. Maximum contaminant levels (MCLs) used to regulate drinking water supplies are based on such a mathematical relationship. Other times the relationships are more complex and require complex equations, such as the description of the movement of a chemical from soil into groundwater below the site and transport of that ground water off-site to some location where it is used as drinking water. Therefore, attributes of the chemical, the landscape, and the people living in the area are all important to understand the dose the people may receive for a given concentration in the soil or landfill.

What is multimedia exposure assessment?

A multimedia, multiple pathway exposure assessment is a mathematical or quantitative relationship between the concentration of a chemical in a waste in a specific location and the daily dose received by people or non-human species. The exposure assessment has three steps.

The first step is to characterize the uptake of a chemical from the environment into the body. Chemical vapors and contaminated dust in air may be inhaled. Contaminated water, soil and food may be eaten and contribute to the total daily dose. Chemicals may be absorbed through the skin from contaminated soil and water in contact with the skin. The exposure media are air, water, soil and food which are inhaled, ingested, or comes into contact with the skin. Mathematical equations relate the concentration in the exposure media to the total daily dose.

Second, mathematical relationships must be established between the concentration of chemical in the exposure media (inhaled air, ingested water, soil that contacts skin, etc.) and the environmental media at

APPENDIX D: DTSC REPORT   199

the landfill. Note that exposure media are distinct from environmental media. Environmental media are the air, ground water, surface water and soil which are located above, below or adjacent to the waste. Nearby residents do not breath air which is directly above the landfill or have direct contact with landfill constituents. Therefore, it is important to establish mathematical relationships between the environmental and exposure concentrations for all relevant media by using a multimedia exposure assessment model.

Third, chemicals in the waste may move upward into the air, downward into the groundwater, partition into plants growing near the waste or migrate into nearby surface water. Therefore, equations are needed that are consistent with known scientific principles to establish a relationship between the chemical in the landfill waste and several of the environmental media in the vicinity of the landfill.

Mathematical Models and Regulatory Decisions

The uncertainties associated with risk assessment are enormous. Some people are uncomfortable with quantitative risk assessment because of this uncertainty and they feel that mathematical models suggest a degree of precision that does not exist. We believe that mathematical models can deal with uncertainty quantitatively and, therefore, their use need not imply an absence of uncertainty. Furthermore, we believe that models are essential for rational decision-making for environmental regulatory agencies. CalTOX and the PEA spreadsheets are quantitative statements about the Department's understanding in how chemicals move in the environment and cause harm to human beings. The equations in those spreadsheets are not proven facts. They must be viewed as hypotheses to update and correct as new information becomes available. The scientific process will inevitably add to our understanding of these processes, but the regulatory agencies must make decisions today. Regulatory agencies must unambiguously lay out the basis for classifying wastes. Equations like those in CalTOX and the PEA provide such an unambiguous method. It is vital that such models be viewed as tools in need of constant maintenance, because the scientific information available to decision-makers in the future will be different than it is today.

How is uncertainty treated in developing TTLCs?

Quantitative risk assessment is based on mathematical equations which relate source concentration to risk. These equations have many variables related to chemical properties, landscape attributes and human exposure parameters. Usually, the values for these variables are not known precisely. These TTLCs are going to be used throughout California. A single value for rainfall or wind speed cannot represent the wide range of rainfalls and wind speeds for all of California. Not all people will live in a house located near a landfill throughout their lives. Some people move frequently, other people stay in a single house for a long time. Since the TTLC is calculated from all these variables, a range of TTLC values is more appropriate than a single arbitrarily chosen value. Therefore, a distribution or range of values was selected for each input parameter. This distribution reflects the diversity of values for landfills found throughout California. The uncertainty and variability reflected by these distributions is propagated through the mathematical equations using a computer-based method called Monte Carlo simulation.

The result of these computer simulations is a distribution of TTLC values which reflects the uncertainty and variability in the underlying input parameters. The TTLCs in all tables are the tenth percentile of those distributions. This indicates that theoretically there is a ten percent chance that an individual living on or near the waste would have exposures exceeding the "acceptable" dose. This does **not** mean that ten percent of these people are at risk of dying of cancer or some other disease. The toxicity criteria used to establish the maximum daily dose do not treat uncertainty with distributions. They are selected to protect everyone given enormous uncertainties and are highly likely to be large over-estimates the true risk. In some cases, the true risk at doses of regulatory interest may be zero. Therefore, the probability of an individual experiencing a health effect from the chemical exposure is actually very much less.

Description of the Exposure Scenarios

In risk assessment, the combination of a source of chemical and population potentially exposed to that chemical are called an exposure scenario. Exposure scenarios must be identified to compute a risk-based TTLCs. For waste classification, the source of the chemical is determined by the disposal of the waste containing that chemical. The potentially exposed populations are the people or non-human species that may be exposed to chemicals in the waste so disposed. The assumed disposal of wastes and potentially exposed-populations are described below.

Disposal Practices

Two distinct disposal practices are assumed for the two levels of TTLCs: disposal in a solid waste landfill and land application. The upper TTLC is used to distinguish waste that can be disposed in a solid waste landfill vs. waste that must be disposed of in a hazardous waste landfill. Therefore, risks associated with solid waste landfill disposal are an appropriate basis for establishing risk-based upper TTLC levels.

The lower TTLC is used to distinguish waste that need not be regulated by DTSC from waste regulated by DTSC. The actual disposition of wastes in California is varied. Significant efforts are underway to divert wastes from landfill to recycling and alternative beneficial uses. Sewage sludge and ashes containing trace elements can be valuable soil amendments. Soil amendments are applied and tilled into land periodically. Therefore, risk associated with land application of waste was selected as the basis for establishing risk-based exit TTLC levels.

Potentially Exposed Populations

Once waste is disposed of in a landfill or applied to land, many different populations of people may be exposed to chemicals in the waste. Populations may be residents, industrial workers, commercial workers, people engaged in recreation, school children, etc. These populations are defined because their location may lead to exposure to chemicals in the disposed waste. The assessed risks are related to the total dose of chemical resulting from chronic or long-term exposures. Risk is often assessed only for the most exposed population because other people would be at less risk than the most exposed group. In general, residents living on or near sources of chemicals experience the highest chronic exposures. Therefore, a residential population was assumed for both the upper and lower TTLC disposal practices.

A number of different factors contribute to greater chronic exposure often experienced by residents as compared with other populations. Residents spend more time at their homes than other groups spend in a single location. This includes more hours each day, more days per year and more years total. Therefore, residents breath more potentially contaminated air than do other populations. In addition, residents engage in activities other populations do not which increase the number of pathways by which the chemical can theoretically travel from the disposed waste to the person. Residents can have gardens with contaminated soil in which food is grown. Gardening and yard work can lead to potentially contaminated soil in contact with the skin. Children inadvertently or deliberately ingesting soil are most likely to do so at home. The majority of tap water is likely to be consumed at home. Residents contact four different exposure media (air, water, food and soil) to a greater degree than any other group. Generally, if residents are protected and the concentrations in these media are the same for all exposed populations, all people are protected.

The media concentrations are not the same for all populations; often concentrations in workplace air and soil are much higher concentrations than at residents. Workers directly in the vicinity of the waste are likely to experience much higher air and skin contact concentration than residents living near a landfill. The workers have fewer pathways and less time of exposure, but higher concentrations. Without conducting a quantitative analysis, it is not possible to know if a worker is

APPENDIX D: DTSC REPORT   201

more or less exposed than a nearby resident.

Finally, non-human species may experience the highest risk if they are particularly sensitive to the effects of the contaminant or if due to their natural history, they are more highly exposed. Therefore, if exposure can occur ecosystem effects must be examined.

The following table shows the four exposure scenarios.

Disposal Practice	Residential Population	Non-Residential Population
Solid Waste Landfill (upper TTLC)	Houses near a landfill	Workers handling waste
Land Application (exit TTLC)	Houses built on land to which waste had been applied	Ecological effects

The non-resident exposure scenarios differ between both disposal practices. Ecological effects were not evaluated for landfill disposal because those effects are not considered to be significantly different than those at hazardous waste landfill. Therefore, it is not rational to require waste to go to a hazardous waste landfill based on ecological effects, so upper TTLCs should not be based on ecological effects. A worker scenario was not explicitly evaluated for the land application disposal practice, because these workers would be less frequently exposed than would the waste worker in the landfill disposal scenario. All people are protected at least to the level of waste workers because of a policy decision resulting from the comparison of proposed exit and upper TTLCs. If the proposed exit TTLC exceeds the proposed upper TTLC, than no exit TTLC is proposed and no special waste category can be established for a given chemical.

Exposure Scenarios

Each of the four exposure scenarios has specific features. These features include the pathways by which a chemical moves from the disposed waste to the potentially exposed population and attributes of the exposed population.

1. Residents Near a Landfill

The upper TTLC residential scenario is people living 100 meters from a landfill accepting wastes containing the TTLC chemicals. Up to 100% of the refuse in the landfill is assumed to contain the TTLC chemical. Landfill area and depth are based on a distribution of these values taken from the Waste Management Unit Database System maintained by the State Water Resources Control Board. Four properties are needed define landfill refuse: fraction of organic carbon, water content, porosity and density. Distributions of values for these properties were taken from the literature. These values may be changed for a specific landfill with verifiable information for the purpose of a variance, but these policy and literature values will be used to compute statewide levels.

Chemicals in waste are transported from the landfill to the residents in the air either as a vapor or adhering to dust particles. The residents may inhale the chemical or it may be deposited onto the soil in the yard. People may come into contact with the contaminated soil. Depending on the chemical, it may be taken up into various foods (including meat, milk, eggs and vegetables) that are consumed by the residents. For some organic chemicals, significant amounts of the chemical can be transferred from a mother's blood into her milk and this leads to high doses to breast-feeding infants.

2. Waste Workers

The other upper TTLC exposure scenario is designed to assess the risk of workers handling the waste. These workers are assumed to handle undiluted waste either before arrival at the landfill or at the landfill. The pathways include inhalation of dusts and/or vapors, inadvertent ingestion of small amounts of waste and uptake of the waste through skin.

3. Resident on Converted Land

A policy decision was made to replicate the assumptions of land application used by U.S. EPA in deriving the maximum concentrations allowed in biosolids (CFR Part 503). These assumptions include an estimate of 0.7 kg waste per m^2 of land for 20 years. After the last application, residential homes are built on the land. The residents have attributes identical to those described above (resident near the landfill) except that they also eat fish from a surface water body adjacent to the land onto which the chemical was applied.

4. Ecosystem Effects

Protection of ecosystems was accomplished through a tiered method. The first tier relied on the risk assessments conducted by the U.S. EPA in support of the Hazardous Waste Identification Rule (HWIR). HWIR is an effort similar to the setting of TTLCs. The objective is to define a concentration in waste that could be used to determine a regulated waste from an unregulated waste. HWIR has computed concentrations for most of the TTLC chemicals based on residential exposure and a variety of ecological effects. Most of the residential based levels are protective of ecological effects, therefore, no further work was done on those chemicals. Ecologically based TTLCs were derived for those chemicals that were driven by ecological concerns in the HWIR studies.

Models used for determining risk-based criteria for the scenarios

Four basic models or approaches were used in determining the concentrations for the four different scenarios shown in Figures 1 and 2. Different models were used for organic chemicals, inorganic lead and all other inorganic chemicals. Table 1 shows the the method name for each chemical-scenario combination.

Table 1: Computational Models Used for the Twelve Scenario-Chemical Combinations

Criteria	Upper TTLC Calculations		Lower TTLC Calculations	
Scenario Modeled	Residents near landfill	Waste workers	Residents on converted land	Ecological concerns
Organic Chemicals	CalTOX Landfill	PEA Worker Organic	CalTOX Land Conversion	Multi-tiered Process
Inorganic Lead	LeadSpread Off-site	LeadSpread Worker	LeadSpread Land Conversion	Multi-tiered Process
Inorganic Chemicals	PEA Off-site	PEA Worker Inorganic	PEA Land Conversion	Multi-tiered Process

Modified versions of the CalTOX model were used for the residential scenarios for organic chemicals for both the upper and lower TTLC. Modified versions of the LeadSpread model were used for all human

APPENDIX D: DTSC REPORT

exposure scenarios to inorganic lead. Modified versions of the Preliminary Endangerment Assessment (PEA) model were used for all human exposures to inorganic chemicals other than lead and waste worker exposure to organic chemicals. The ecological effects for all chemicals were evaluated by using a screening process to identify chemicals for which ecological effects occurred at lower doses than for human health effects.

What is CalTOX and why was it chosen?

The CalTOX risk assessment framework is a comprehensive multiple pathway, multimedia approach to relating risk and concentration of chemical in soil. This model is an extension of the current US EPA risk assessment policy defined in the Risk Assessment Guidance for Superfund (RAGS) series. The model represents the state of the art in modeling multimedia distribution of environmental contaminants and multi-pathway exposures of human populations. Among the distinguishing features of CalTOX acknowledged by its reviewers are explicit treatment of mass conservation and chemical equilibrium, calculation of gains and losses in multiple environmental media compartments (air, soil, groundwater, etc.) over time by accounting for both transport among the compartments and transformation within compartments, and the quantitative and comprehensive treatment of uncertainty and variability. The model for on-site exposure (when the people live on contaminated soil) has been extensively peer-reviewed and received numerous commendations. US EPA's scientific advisory panel referred to it as "...potentially the most advanced of all of the models reviewed with respect to exposure..." in a report on Human Exposure Assessment. The CalEPA Risk Assessment Advisory Committee describes the model favorably. During its development, comments were solicited from internationally known scientists and are summarized in Part IV of the technical document describing the model. CalTOX is recognized as a leading model for the kind of approach it implements in multimedia risk assessment.

This model is shown in schematic form in Figure 1. An initial chemical concentration is specified for either the root zone soil or the vadose zone soil. Over time the chemical flows in the direction of the arrows, the rate at which it flows for any given arrow is dependent on the properties of the chemical and the environment. The model is divided into three groups of equations: exposure assessment based on RAGS, multimedia transport and intermedia transfer. The RAGS exposure equations represented on the right of the diagram defines how much chemical people take into their body from four contaminated exposure media: food, air, water and soil. The multimedia transport equations represented on the left predicts the concentration in seven environmental compartments at various points in time. The intermedia transfer represented in between relate the concentration in exposure media to concentrations in the environmental compartments.

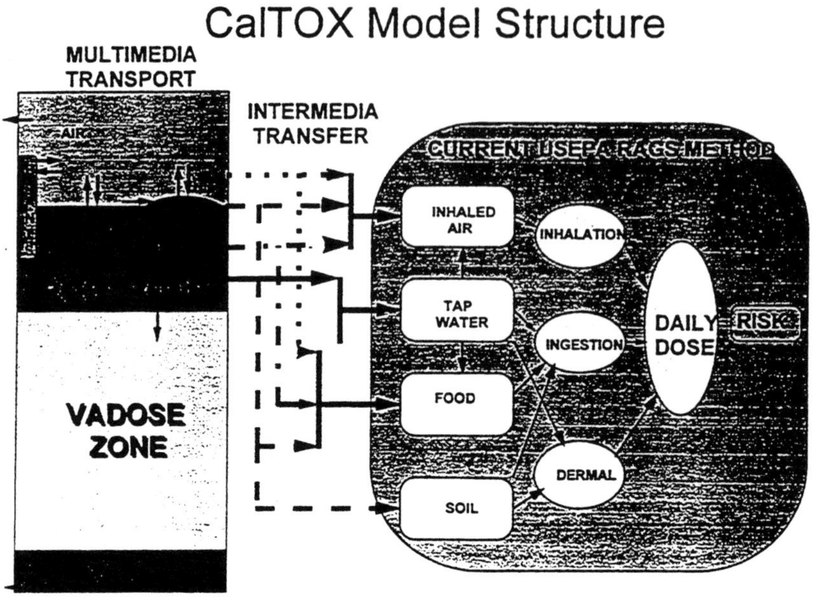

CalTOX Model Structure

Figure 1: Diagram of the CalTOX model

What modifications have been made to CalTOX and why?

The current version of CalTOX, CalTOX 2.3, is designed to assess risk to people living on or nearby soil containing a fixed concentration of chemical. CalTOX 2.3 is not designed to model risks to people living near a landfill or continuous addition to soil followed by occupation by residents. Therefore, two modified versions of CalTOX 2.3 were created to model the landfill and converted land scenarios. Both modifications required changes to the equations in the model as well as changes to the default input parameters.

CalTOX *Landfill* was created by transforming the root zone compartment of CalTOX 2.3 into a landfill compartment. This involved two changes to the model structure and several different input variable values. The first change in the model was to relate the landfill compartment concentration to the waste concentration by allowing the use of a waste dilution factor. The second change was to add an estimate of transport of chemical in gases produced in the landfill (carbon dioxide and methane) from the landfill compartment to the air. Landfills have different areas and depths than residential yards. Landfill contents have different properties than soil. Therefore, mean estimates of those parameters were selected based on literature values. Neither the waste dilution factor nor the distance off-site were treated stochastically. These were viewed as policy decisions. The waste dilution factor chosen was one, and residents were located 100 meters off-site.

CalTOX *Land Conversion* was created by computing the root soil concentration from an application rate, mixing depth, application duration and waste concentration rather than specifying an initial root soil concentration. The values for these new parameters were the point estimates cited in the US EPA technical background document used in promulgating the regulations for biosolid application to land

APPENDIX D: DTSC REPORT

(40CFR Part 503.13). Therefore, like CalTOX *Landfill*, a relationship is established between health risk and waste concentration rather than root soil concentration.

Detailed descriptions of the exact changes to CalTOX 2.3 are described in the report *CalTOX Adaptations for Derivation of Exit and Upper TTLC Criteria* located in Tab 4a of the NAS notebook. The two modified spreadsheets were used to compute distributions of waste concentrations for each of the organic chemicals. The 10th percentile of each of these distributions was identified and presented in Table 3 shown below.

CalTOX was not used to model inorganic chemicals, including lead, because estimates of the soil-water partition coefficient (K_d) is required for each chemical. For organic chemicals, the K_d can be estimated from chemical properties and the properties of the soil. Such estimates are not possible with inorganic chemicals because their interaction with soil is more complicated. Therefore, models other than CalTOX were required for inorganic chemicals.

What is the Lead spreadsheet model and why was it chosen?

More data are available on source concentrations, like soil, and human blood lead concentrations of lead than any other chemical. These data have been used to develop a model for predicting a relationship of the soil concentration of lead and potential health effects. This model is implemented as a Department of Toxic Substances Control Lead Risk Assessment Spreadsheet (LeadSpread). Like CalTOX, LeadSpread was designed for hazardous waste sites, and it has been used by the Department for evaluating the risk associated with contaminated soil for approximately six years. Slight modifications were needed to adapt the model for use in establishing human-health-based candidate TTLCs for lead. The model and its modifications for this application are described in more detail, and example spreadsheets can be found under Tab 4b.

What modifications have been made to the Lead spreadsheet model and why?

- LeadSpread Off-site was used to compute the risk to residents living 100 meters from a landfill containing lead wastes. For the offsite version used in the calculation of upper TTLCs for nearby residents, a soil dilution factor of 0.00098 is incorporated to estimate the concentration in off-site soil from the concentration in the waste. This value was derived from calculations for other inorganics using the PEA model (tab 4c), and affects the soil ingestion, plant ingestion, and dermal exposure pathways.
- LeadSpread Land Conversion was used to compute the risk to residents moving into homes built on soil into which waste had been mixed.
- LeadSpread Worker was used to to compute the risk to workers handling the waste.

What is the Preliminary Endangerment Assessment model and why was it chosen?

CalTOX was not appropriate to use for estimating risks and hazards for the worker scenario on-site for organic chemicals, because the fate and transport capability of CalTOX was not needed, and because worker exposure does not include the indirect pathways of consumption of meat, milk, eggs, and homegrown produce. CalTOX also lacks input distributions for important fate and transport parameters for inorganics. DTSC (1994) published a model useful for these scenarios in its "Preliminary Endangerment Assessment Guidance Manual" (PEA). The PEA model was successfully adapted to estimate TTLCs for inorganic constituents in wastes and for worker-based TTLCs for organics.

The PEA is an adaptation of the methods recommended by USEPA (1989) in "Risk Assessment for Superfund" (RAGS). The PEA method has been in use in DTSC's Site Mitigation Program for several years. Since both CalTOX and the PEA model are based on RAGS, the intake equations for exposure and the toxicity criteria are identical in the PEA and CalTOX. However, CalTOX considers additional pathways not considered in the PEA equations.

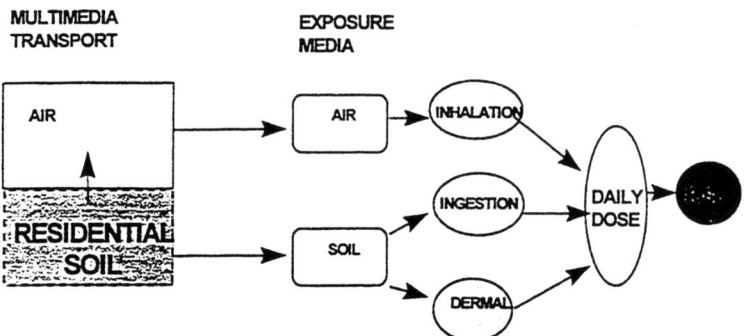

Figure 2. Preliminary Endangerment Assessment Model

What modifications have been made to the PEA and why?

Like CalTOX and Leadspread, the PEA was designed for hazardous waste sites. Since none of the applications of PEA in calculating TTLCs is identical to a hazardous waste site, some modifications had to be made. The PEA method was translated into spreadsheet form to estimate TTLCs for three scenarios: exposures to organic and inorganic chemicals for onsite hazardous waste workers, exposure to inorganic chemicals for a resident near a landfill, and exposure to inorganic chemicals for a resident on converted land (Table 1).

1. On-Site Hazardous Waste Worker

This differed from the original PEA model in two ways. First, the receptor is a worker, whereas the original PEA used a residential setting. Second, the landscape is a landfill or waste handling facility, whereas the original PEA used a residential landscape. Two spreadsheets were constructed, entitled "PEA Worker Organic" and PEA Worker Inorganic". Because of the necessity to model vapor emissions for organic chemicals, the spreadsheet model for organics is considerably more complex than the one for inorganic constituents. The PEA on-site worker version models only inhalation of vapors and dust, dermal contact with soil, and ingestion of soil.

2. Resident Near a Landfill

This exposure setting was similar to the original PEA in that it is residential, but it differs in that the source of toxicants is dust blowing from the landfill or other waste facility to settle on residential soil. The off-site resident was exposed directly to inhaled dust from the facility, but exposures from soil ingestion, and dermal contact with soil included assumptions about dilution of windblown dust into residential soil over a period of years. For convenience, the model was divided into two spreadsheets, "PEA Offsite Hazard", and "PEA Offsite Risk".

3. Resident on Converted Land

This exposure setting is residential, as for the original PEA, but for this scenario, a dilution factor is applied to all four exposure pathways: ingestion of soil, dermal contact with soil, ingestion of homegrown produce, and inhalation of dust. The dilution arises as waste is applied to and mixed

APPENDIX D: DTSC REPORT

with soil over a period of years. Thus, the target concentration in the waste is higher than the concentrations to which the resident will eventually be exposed. It was assumed that this activity will resemble application of sewage sludge to agricultural land, followed by conversion of the land to residential use. For convenience, the model was divided into two spreadsheets, "PEA Land Conversion Hazard", and "PEA Land Conversion Risk".

How were effects on non-human species evaluated?

The candidate lower (exit-level) TTLCs based on human health were reviewed to determine whether or not they would protect non-human species (see Tab 4d for details). Upper TTLCs were not similarly reviewed because protection of ecological resources is thought to be similar in Class 1 versus Class 2 or 3 landfills. A tiered screening approach was adopted due to the lack of an accepted risk assessment method for ecological toxicity. The first screen consisted of a comparison of the human toxicity exit concentration to the ecological toxicity exit for each chemical in the review drafts of the USEPA Hazardous Waste Identification Rule (HWIR, 1). For 29 of the 37 chemicals being analyzed, there were HWIR exit concentrations for both humans and ecosystems. The other eight were considered low priority in the HWIR analysis and therefore are not likely to present significant threats to ecosystems at concentrations below those that would be of concern for human health.

For 18 of the 29 chemicals the human toxicity exit concentration was lower than the ecological toxicity exit concentration, indicating that for those 18 chemicals, human-health-based lower TTLCs would be likely to also protect other species. For hexavalent chromium, the human-health based TTLC is the same as the HWIR ecological exit concentration. The remaining ten chemicals - endrin, methoxychlor, lead, mercury, selenium, nickel, vanadium, cadmium, zinc,and copper - moved to the second screen.

Table 2: HWIR Human-Ecological Risk Comparison

chemical	HWIR human exit level*	HWIR ecological exit level*
Aldrin	0.00009	0.008
Chlordane	0.002	4
DDT & congeners	0.0002	0.004
2,4 D	600	none
Dieldrin	0.0003	0.05
Dioxin	5e-7	1e-6
Endrin	0.6	0.1
Heptachlor	0.8	20
Kepone	0.0003	3
Lead (organic)	none	none
Lindane	0.009	0.2
Methoxychlor	300	3
Mirex	none	none
Pentachlorophenol	0.5	50
Polychlorinated biphenyls	0.0007	0.05
Tricholoethylene (TCE)	10	none
Toxaphene	0.00003	0.001
Silvex	100	400
Vinyl chloride	0.06	none
Antimony	2	4
Arsenic	0.08	10
Barium	2000	4000
Beryllium	0.01	20
Cadmium	10	5
Hexavalent Chromium	10	2**
Cobalt	none	none
Copper	900	1
Fluoride	none	none
Lead	200	1
Mercury	5	0.2
Molybdenum	100	200
Nickel	1000	20
Selenium	8	0.4
Thallium	2	none
Vanadium	300	20
Zinc	8000	0.4

* Lower value shown in **bold**
** Same value as proposed TTLC based on human health

APPENDIX D: DTSC REPORT

The second screen involved a consideration of the soluble or extractable fraction of the ten waste constituents. The limiting ecological endpoint for nickel and copper was toxicity to aquatic plants (1). Since the proposed SERTs for these metals are based on Ambient Water Quality Criteria for the protection of aquatic life, the SERTs should protect aquatic plants. Therefore no changes are proposed in the human-health-based exit-level TTLCs for these elements.

Vanadium The limiting ecological endpoint in the HWIR analysis for vanadium was phytotoxicity. Estimated no-effect concentrations range from 2 to 20 mg/kg (1,3). However, these values are below even minimum ambient levels in California and thus cannot be used as the basis for regulatory standards. The STLC is the basis of the the proposed TTLC for vanadium (see page 37).

Cadmium The draft HWIR ecological exit concentration for cadmium, based on toxicity to soil fauna, is 5 mg/kg (1). The lowest Oak Ridge National Laboratory screening concentration of cadmium, based on phytotoxicity, is 3 mg/kg (3). Assuming that the waste is applied to the land at the rate of 7000 kg/ha/year for 20 years as described previously, the maximum cadmium concentration in the waste that would not result in a final soil comcentration exceeding the 3 mg/kg screening level for phytotoxicity is 60 mg/kg.

Lead The draft HWIR ecological exit concentration for lead is 1 mg/kg, based on toxicity to soil fauna (1). This value cannot be used as a basis for defining hazardous waste in California, where background soils contain a minimum of 14 times that amount of lead. The ORNL screening concentrations based on phytotoxicity and soil fauna toxicity are 50 and 500 mg/kg, respectively (2,3). Assuming that the waste is applied to the land at the rate of 7000 kg/ha/year for 20 years as described previously, the maximum concentration of lead in the waste that would not result in a final soil comcentration exceeding the 50 ppm screening level for phytotoxicity is 990 ppm. Rounded to one significant figure, this results in a suggested exit-level TTLC of 1000 mg/kg.

Zinc The draft HWIR ecological exit concentrations for zinc is 0.4 mg/kg, based on toxicity to soil fauna (1). The ORNL screening levels for zinc are 50 and 100 mg/kg, based on phytotoxicity and soil microorganisms, respectively. These concentrations are well below 90 th percentile background concentrations in the California soils (tab 9), and thus cannot be used as a basis for defining hazardous waste in California. Since no satisfactory criterion is available for total zinc, DTSC proposes to regulate zinc in solid waste only by its soluble or extractable concentration.

Selenium The limiting toxicological endpoint for selenium is reproductive toxicity in waterfowl. The threshold for this endpoint can be estimated in a variety of ways (see Tab 8). The California EPA Office of Environmental Health Hazard Assessment (OEHHA) recommends calculating the endpoint using a benchmark dose and a sediment-fish transfer factor (ranging from 3.9 to15.6), and assuming 100% foraging on a contaminated site as well as conditions favoring selenium uptake into dietary items of aquatic birds. However, in order to be consistent with the approach used throughout this proposed revision of the waste classification system, the approach of Van Derveer and Canton (Tab 8) was selected as the basis for the proposed lower TTLC for selenium. Those authors determined EC_{10} values (10th percentile estimates of the threshold for selenium toxicity in sediment) of 2.5 and 4.0 mg/kg, dry weight, associated with predicted and observed effects, respectively, in wild fish and birds. Assuming that the waste is applied to the land at the rate of 7000 kg/ha/year for 20 years as described previously, the maximum selenium concentration in the waste that would not result in a final soil concentration exceeding the 2.5 mg/kg EC_{10} based on predicted effects for waterfowl toxicity is 50 mg/kg. The proposed ecosystem-based, exit-level TTLC for selenium is therefore 50 mg/kg, dry weight (however, see page 37).

Mercury According to the OEHHA analysis, the limiting pathway for mercury is biomethylation, uptake by benthic organisms, and food web transfer of the organomercury to fish and ultimately to predator fish, mammals, and birds (Tab 8). The belted kingfisher was chosen as the species of concern because it is a sensitive fish-consuming bird, and birds appear to be at least as sensitive to methymercury as fish or mammals. Using the mallard dietary reproductive no-adverse-effect concentration (NOAEL) for methylmercury of 0.05 mg/kg diet, with an interspecies conversion factor of 3.2 (reflecting a daily food consumption rate of 15.6% body weight for the mallard versus 50% body weight for the kingfisher), and a conservative assumption of 100% foraging on contaminated fish, a dietary NOAEL for the kingfisher would be 0.05/3.2 = 0.016 mg/kg diet.

In order to relate the mercury concentration in the kingfisher's diet to a concentration in sediment, a sediment-to-fish transfer factor is needed. A range of sediment-to-fish transfer factors identified from the literature for California (0.013-20; Table 1, Tab 4d). Combining these transfer factors with the estimated dietary NOAEL for the kingfisher, an estimated range of maximum "safe" sediment concentrations for the kingfisher would be 0.0008 to 1.2 mg/kg sediment, dry weight. Assuming that the concentration in the sediment is the same as the concentration in the soil (for example, fields to which the waste has been applied may be flooded to form wetlands) and that the waste is applied to the land at the rate of 7000 kg/ha/year for 20 years as described previously, the maximum mercury concentration in the waste that would not result in a final sediment concentration exceeding the 0.0008 to 1.2 mg/kg sediment, dry weight 0.016 mg/kg screening sediment concentration for food-web toxicity is (0.0008 to 1.2) x 19.8 = 0.011 to 17 mg/kg sediment (geometric mean 0.19).

Endrin The HWIR exit concentration for endrin is 0.1, based on toxicity to the great blue heron via the aquatic food web. This concentration was accepted as the eco-based TTLC for endrin because it was verified using the environmental transport modeling of endrin from waste to fish in CalTOX *Land Conversion*.

Methoxychlor The HWIR exit concentration for methoxychlor is 3, based on toxicity to sediment-dwelling organisms. The HWIR benchmark dose for sedim

ent-dwelling organisms is converted from the ambient water quality criterion for the protection of aquatic life (AWQC) of 0.00003 mg/l, using equilibrium partitioning. This conversion involves the implied assumption that toxicity to benthic organisms can be predicted by pore water concentration in sediment and that these organisms will be protected if pore water concentration does not exceed the AWQC. The same thing is accomplished by limiting the soluble or extractable concentration of endrin in the waste to 100 times the AWQC. The latter is DTSC's proposed approch. Thus, no change is proposed in the human-health-based TTLC of 100.

References

1. U.S. Environmental Protection Agency, 1995, Technical Support Document for the Hasardous Waste Identification Rule: Risk Assessment for Human and Ecological Receptors, Office of Solid Waste Contract # 68-D2-0065, 68-W3-0028.
2. Oak Ridge National Laboratory, undated, Toxicological Benchmarks for Potential Contaminants of Concern for Effects on Soil and Litter Invertebrates and Heterotrophic Process, ES/ER/TM-126/R1.
3. Oak Ridge National Laboratory, 1995, Toxicological Benchmarks for Screening Potential Contaminants of Concern for Effects on Terrestrial Plants: 1995 Revision, ES/ER/TM-85/R2.
4. Wildlife Exposure Factors Handbook. US EPA 1993, EPA/600/R-93/187a Office of Research and Development

APPENDIX D: DTSC REPORT

Table 3: Risk-based Candidate TTLCs

Chemical	Upper TTLC (mg/kg) Nearby Residents	Upper TTLC (mg/kg) Waste workers	Lower TTLC (mg/kg) LCS Residents	Lower TTLC (mg/kg) Ecological concerns
Aldrin	0.006	3	0.0009	protected[e]
Chlordane	1	30	0.06	protected[e]
DDD	20	200	0.3	protected[e]
DDE	1	150	0.3	protected[e]
DDT	4	100	0.7	protected[e]
2,4-Dichlorophenoxyacetic acid[f]	3,000	1500	50,000[d]	protected[f]
Dieldrin	0.2	3	0.004	protected[e]
Endrin	70	150	0.8	0.1
Heptachlor	0.7	6	2[d]	protected[e]
Kepone	0.2	3	0.02	protected[e]
Tetraethyl Lead	8×10^{-6}	0.0003	0.0007^{d}	protected[f]
Lindane	60	30	5	protected[e]
Methoxychlor	7,000	2,000	100	6,400
Mirex	0.9	2	0.04	protected[f]
Pentachlorophenol[i]	500	500	400	protected[e]
Polychlorinated biphenyls (PCBs)[a]	nd	nd	nd	nd
Tricholoethylene (TCE)	20	70	2,000[d]	protected[e]
Toxaphene	0.04	30	6[d]	protected[e]
2,4,5-Trichlorophenoxyproprionic acid[i]	2,000	1000	40,000[d]	protected[e]
Vinyl chloride	0.2	0.7	50[d]	protected[f]
2,3,7,8-Tetrachlorodibenzodioxin	7×10^{-7}	5×10^{-4}	1×10^{-7}	protected[e]
Inorganic lead	20,000	6,000	5,000	1000
Antimony	15,000	700	4,000[d]	protected[e]
Arsenic	200	40	400[d]	protected[e]
Asbestos[b]	nd	nd	nd	nd
Barium (excluding barite)	>1,000,000	100,000	700,000[d]	protected[e]
Beryllium	300	20	200[d]	protected[e]
Cadmium	150	500	3,000	60
Trivalent Chromium[c]	>1,000,000	>1,000,000	>1,000,000	protected[e]
Hexavalent Chromium	5	15	80[d]	protected[f]
Cobalt	15,000	20,000	200,000[d]	protected[e]
Copper	>1,000,000	70,000	400,000[d]	protected[g]
Fluoride	>1,000,000	100,000	600,000[d]	protected[e]
Ionic Mercury	10,000	500	3,000	0.2
Molybdenum	200,000	9,000	50,000[d]	protected[e]
Nickel	3,000	7,000	50,000[d]	protected[g]
Selenium	200,000	9,000	50,000	40
Thallium	3,000	150	800[d]	protected[e]
Vanadium	300,000	10,000	70,000	2-20
Zinc	>1,000,000	500,000	>1,000,000[d]	protected[h]

ᵃ New TTLCs for PCBs have not been determined.

ᵇ New TTLCs for asbestos have not been computed.

ᶜ For these scenarios, CrIII does not pose a health threat.

ᵈ The lower TTLC values were greater than the upper TTLC.

ᵉ Using the HWIR ratio, the residential TTLC protects ecosystems.

ᶠ Chemical not considered by EPA to be a priority for eco-assessment.

ᵍ The SERTs protect against toxicity to aquatic plants.

ʰ No acceptable eco-based concentration available. Zinc in waste will be limited by their soluble or extractable fraction only.

ⁱ Chemical parameters for the unionized moiety were used.

ʲ Mercury range 0.011-17 with geometric mean of 0.19 mg/kg.(see text)

The spreadsheets named in Table 1 and described in the sections above were used to compute risk-based concentrations for the four scenarios. The results of these computations are presented in the four columns of Table 3. The rows of Table 3 are divided with the organic chemicals (CalTOX) shown in the top half, inorganic lead (Lead spreadsheet) shown as a single line in the center, and other inorganic chemicals (PEA) shown in the bottom half.

The list of chemicals for which TTLCs were computed is based on the list of chemicals for which TTLCs currently exist in regulation. The chemical names in Table 3 differ than those currently appearing in regulation for several chemicals. First, DDE DDT and DDD all appear separately in a box in the table. DDT can be converted into DDE or DDD. Therefore, risk-based TTLC values were computed for all three chemicals. The lowest of the six upper values (DDE-nearby resident) was selected as the risk-based upper-level TTLC. The lowest of the three land conversion residents (DDE) was selected as the lower risk-based TTLC. Second, risk-based computations require that a single chemical with specific chemical and toxicological properties be identified to compute a TTLC. This can lead to identifying a single chemical as a surrogate for other similar chemicals. Tetraethyl lead is used as a surrogate for all organic lead. Third, vinyl chloride has been moved from to the list of carcinogens in section 66261.24 of the current regulation to this list because of the importance of this chemical. Fourth, total chromium has been replaced by trivalent chromium. All chromium is either considered trivalent or hexavalent because of the large differences in the toxicity of two valances. Fifth, mercury has been restricted to ionic mercury excluding organic mercury and elemental mercury. Sixth, silver has been eliminated from the list because the form found in the environment is known not to be bioavailable.

There are also several explanations required of values in the Table 3. First, no new TTLCs are being proposed for PCBs or asbestos. Both the analytical methods and toxicity criteria for PCBs are in a state of change currently. Efforts are under way to create a toxicity equivalency factor approach for PCBs similar to dioxin. Therefore the department is waiting for the outcome of these efforts before proposing a new TTLC. Asbestos risks are based on fiber counts. This information has not changed greatly, so an update was deemed unnecessary. Second, the organic acids pentachlorophenol, 2,4-dichlorophenoxyacetic acid (2,4-D) and 2,4,5-trichlorophenoxyproprionic acid (2,4,5-T) exist in an unionzed form at lower pHs. This form is more volatile and less water soluble than the ionized form. Since the objective of the TTLCs is to model the exposure pathways other than those involving ground water, the chemical characteristics of the unionized form were used to predict the fate of these ground chemicals in the environment. At pH values between 5 and 9, the environmental fate of pentachlorophenol will be well represented by the CalTOX equations; prediction errors are larger for 2,4-D and 2,4,5-T.

APPENDIX D: DTSC REPORT 213

Second Stage: Detection Limits and Ambient Concentrations

Following the calculation of risk-based candidate TTLCs, these values are evaluated to determine whether they can be practically implemented as regulatory limits. First, a policy decision was made not to set any TTLC lower than could be measured, since a toxicity threshold is not a useful criterion if it is so low that it cannot be measured. Second, a policy decision was made not to set any TTLC lower than concentrations which are naturally present in California soils or widely distributed in the environment.

Estimated Quantification Limits

Estimated quantitation limits (EQLs) are defined as the lowest concentration that can be reliably achieved within specified limits of precision and accuracy during routine laboratory operating conditions (Tab 5a). The EQL considers the limitations of the analytical method and the effects of processing the sample matrix. The EQL for substances in complex matrices, such as oily sludges, can be quite high. An EQL is calculated for each chemical for each regulatory limit class (upper TTLC, lower TTLC). Each EQL is then multiplied by two because in order to statistically evaluate compliance with a standard, one must be able to measure concentrations above and below the standard.

Risk-based candidate TTLC which were less than twice the EQL were changed to twice the EQL. The proposed upper TTLC for toxaphene and proposed lower TTLCs for chlordane, heptachlor, methoxychlor, toxaphene, Silvex, and vinyl chloride are based on EQLs.

Comparison with Background Concentrations

Inorganics: A two-step process was used to implement the policy decision to consider background concentrations in setting TTLCs for inorganic chemicals:
(1) All calculated health-based levels for inorganic chemicals regulated by DTSC (except fluoride and hexavalent chromium), were compared with maximum background levels found in native California soils, as reported in the University of California, Riverside study. The risk-based concentrations for mercury and vanadium were less than their maximum background concentrations.
(2) Determine the concentrations of vanadium and mercury in waste that would not cause a significant increase in background concentrations of these substances, when the wastes were mixed with soil as postulated in the land conversion scenario, discussed earlier. The following table shows the calculation steps for vanadium and mercury.

	UCR soils data			USGS soils data			average	waste
	mean	90%	difference	mean	90%	difference	difference	concentration
mercury	0.26	0.612	0.352	0.154	0.47	0.316	0.3342	6.62
vanadium	24.3	185.2	161	124.8	200	75.2	118	2337

A "significant increase" was defined as the difference (cloumns 4 & 7) between the means (columns 2 & 5) and the ninetieth percentiles (columns 3 & 6). This calculation was done for the UCR data (1) and the USGS data (2), and the the results averaged (column 8). Since the land conversion scenario results in a 19.8-fold dilution of the waste as it is mixed with soil, the equivalent waste concentration is 19.8 times this average difference (column 9). This means that if waste containing mercury or vanadium at the concentration in column 9 is mixed with soil that contains average background concentrations of these metals, the resuliting mix will contain no more mercury or vanadium than the 90th percentile of background.

Organics: Dioxins and dibenzofurans are generally formed as a result of human activities. As sources of these compounds are brought under control, the levels in the environment are dropping and will probably continue to do so. Published data on the distribution of dioxins in the environment in the U.S. and the U.K. were used to estimate ambient concentrations in soils, which appear to have a mean and standard deviation of approximately 0.000008 mg/kg (TEQ). Second, the concentration of dioxin TEF in waste that would not cause a significant increase in background concentrations of these substances was determined. A "significant increase" in this case, was defined as an increase of 1.28 standard devations, or 0.00001 mg/kg (TEQ). The concentration in waste that would produce such change in soil, estimated using the land conversion scenario, was 0.0002 mg/kg (0.00001 x 19.8). For each chemical, the value that is the basis for the TTLC is shown in bold in Table 4:

APPENDIX D: DTSC REPORT

Table 4: Comparison of Risk-based Levels with Quantitation Limits and Ambient Levels

Chemical	Upper TT (mg/kg)			Lower TT (mg/kg)		
	Risk-based Level	Estimated Quantitation Limit	Ambient Level	Risk-based Level	Estimated Quantitation Limit	Ambient Level
Aldrin	0.006	0.68	na	0.0009	0.046	na
Chlordane	1	0.74	na	0.06	0.05	na
DDT & congeners	1	0.5	na	0.3	0.034	na
2,4 D	3000	4	na	none[d]		
Dieldrin	0.2	0.88	na	0.004	0.059	na
Endrin	70	0.78	na	0.1	0.052	na
Heptachlor	0.7	0.8	na	none[d]		
Kepone	0.2	40	na	0.02	2.7	na
Organic Lead	8×10^{-6}	100	na	8×10^{-6f}	10	na
Lindane	30	0.5	na	5	0.034	na
Methoxychlor	2000	1.7	na	100	0.12	na
Mirex	0.9	0.3	na	0.04	0.02	na
Pentachlorophenol	500	1.5	na	400	0.1	na
Polychlorinated biphenyls[a]	nd	nd	nd	nd	nd	nd
Tricholoethylene (TCE)	20	1.2	na	none[d]		
Toxaphene	0.04	1.7	na	0.04[f]	0.1	
2,4,5-T	2000	1.5	na	none[d]		
Vinyl chloride	0.2	1.2	na	0.2[f]	0.01	
PCDD/PCDF (TEQs)[b]	7×10^{-7}	See comment[b]	2×10^{-4}	1×10^{-7}	See comment[b]	2×10^{-4}
Inorganic lead	6000	6	97.1	700	0.6	97.1
Antimony	700	120	1.95	none[d]		
Arsenic	50	20	11	none[d]		
Asbestos[c]	nd	nd	nd	nd	nd	nd
Barium (excluding barite)	100,000	400	1400	none[d]		
Beryllium	30	10	2.7	none[d]		
Cadmium	200	10	1.7	40	1	1.7
Hexavalent Chromium	5	2	na	none[d]		
Cobalt	10,000	100	46.9	none[d]		
Copper	60,000	50	96.4	none[d]		
Fluoride	100,000	100	na	none[d]		
Ionic Mercury	500	0.4	0.54	0.2	0.04	7[e]
Molybdenum	9,000	100	9.6	none[d]		
Nickel	3,000	80	509	none[d]		
Selenium	9,000	10	0.43	40	1	3.3[e]
Thallium	100	20	36.2	none[d]		
Vanadium	10,000	100	190	20	10	2000[e]
Zinc	500,000	40	236	none[d]		

Human and Ecological Risk Division   2/27/98

a New TTLCs for PCBs were not determined. See text.
b This is to be applied to dioxin TEQs not a single congener
c New TTLCs for asbestos have not been computed and the existing TTLCs will be used.
d The lower TTLC values were greater than the upper TTLC, therefore, no lower TTLC is proposed for these chemicals.
e The maximum waste concentration based on background considerations.
f These are the upper TTLC values.

Four signficant differences can be seen in the chemical names listed in Table 4 as compared with Table 3. *(1)* The three chemicals DDT, DDE and DDD have been reduced to a single line called DDT and congeners. The lowest risk-based concentration of the three congeners has been selected because DDT can be transformed into DDE or DDD. *(2)* The criterion for dioxins applies to 11 dioxin congeners, not just to 2,3,7,8 tetrachlorodibenzodioxin. This is based on the toxicity equivalence factors (TEFs) used by US EPA. Therefore, waste being classified according to their content of dioxins will need to be analyzed for these 11 congeners. The 11 TEFs will be used to compute an equivalent concentration of TCDD. *(3)* tetraethyl lead has been replaced with organic lead because tetraethyl lead was a surrogate for organic lead. *(4)* Since non-extractable chromium III and zinc were found to pose no significant effect on human health and the environment, the department proposes to regulate those chemicals only by their SERTs.

References

1. Kearney Foundation of Soil Science, Division of Agriculture andf Natural Resources, University of California, Riverside, 1996, Background Concentrations of Trace and Major Elements in California Soils.

2. Boerngen, J.G. and H.T. Shacklette (1981) Chemical analysis of soils and other surficial materials of the coterminous United States. Open-file Report 81-197, U.S. Department of Interior, Geological Survey.

APPENDIX D: DTSC REPORT   217

Appendix 4. Acute Toxicity Thresholds

Oral LD_{50}

DTSC has developed recommended acute oral toxicity thresholds for hazardous wastes and special wastes. Hazardous wastes would include wastes with an oral LD_{50} less than 30 mg/kg. Non-hazardous wastes would include those with an oral LD_{50} exceeding 500 mg/kg. Special wastes would include wastes with oral LD_{50}s between 30 and 500 mg/kg. These thresholds are calculated as follows:

The Hazardous waste threshold is based on an adult exposure scenario because Special Wastes would be accessible to adults but not ordinarily be accessible to children. Adults are assumed to ingest 0.31 mg of waste per kg body weight., The Special waste threshold is based on a child exposure scenario because in order to be unregulated by the Department, wastes should not be an acute toxicity threat to children. Children are assumed to ingest 5 mg of waste per kg body weight.. The waste ingestion rates for adults and children are 90th percentile estimates of inadvertent soil ingestion derived from the CalTOX model. The means (and coefficients of variation) of those distributions are 1.4e-7 (2) and 2.2e-6 (3), respectively. Uncertainty factors of ten to account for the use of laboratory animal toxicity data to predict human toxicity and ten to extrapolate from a lethal concentration to a minimal-effect concentration were multiplied by the waste ingestion rates to arrive at the acute oral toxicity thresholds.

Dermal LD_{50}

DTSC has developed recommended acute dermal toxicity thresholds for hazardous wastes and special wastes. Hazardous wastes would include wastes with a dermal LD_{50} less than 5500 mg/kg. Non-hazardous wastes would include those with an oral LD_{50} exceeding 7400 mg/kg. Special wastes would include wastes with oral LD_{50}s between 5500 and 7400 mg/kg. These thresholds are calculated as follows:

The hazardous waste threshold is based on a dermal contact rate of 55 mg of waste per kg body weight per day by an adult, with an uncertainty factor of 100. The special waste threshold is based on a dermal contact rate of 74 mg of waste per kg body weight per day by a child, with an uncertainty factor of 100. The dermal contact rates for adults and children are 90th percentile estimates of inadvertent contamination of skin by waste using exposure parameters from the CalTOX model. The parameters include fraction of skin surface exposed (mean 30%, SD 1%), soil adhesion (mean 0.5 mg/cm^2, SD 0.2), and skin surface area (adult mean 0.024 m^2/kg, SD 0.001, child mean 0.032 m^2/kg, SD 0.003). Uncertainty factors of ten to account for the use of laboratory animal toxicity data to predict human toxicity and ten to extrapolate from a lethal concentration to a minimal-effect concentration were multiplied by the dermal contact rates to arrive at the acute dermal toxicity thresholds.

Inhalation LC_{50}

The purpose of the proposed revisions in the acute inhalation toxicity threshold is to take into account potential exposure as well as toxicity. The proposal would subdivide the regulated wastes into two categories based on the severity of the threat in order to avoid over-regulating wastes which have low exposure potential (low volatility and limited respirability). The waste being classified would not need to be tested if standard reference values are are available for its toxic constituents.

<u>Volatiles:</u> In order to account for both a chemical's acute inhalation toxicity and its tendency to vaporize, classification of a waste containing volatile constituents would be based on the ratio of each constituent chemical's vapor pressure (in ppm @ 250 C) to its inhalation LC_{50} (in ppm). If this ratio exceeds 0.1, the waste containing the chemical would be a special waste. If this ratio exceeds 1, the waste containing the chemical would be a hazardous waste. These ratios must be summed for wastes with multiple volatile chemicals, i.e. $\Sigma(VP/LC_{50}) > 0.1$ yields a special waste classification and $\Sigma(VP/LC_{50}) > 1$ yields a hazardous classification. Vapor pressure in mm Hg is converted to vapor pressure in atmospheres by dividing by 760. This, in turn is converted to ppm by multiplying by 1 million. The rationale for the proposed thresholds is as follows:

A chemical's vapor pressure in atmospheres multiplied by 1 million gives its theoretical maximum concentration in a closed space in ppm, i.e. concentration = $10^6 * V_p$ (in mm Hg) / 760, which can be simplified to $V_P / 0.00076$. If this concentration exceeds the LC_{50} for a volatile chemical, then the chemical could form a lethal atmosphere, and is therefore considered a hazardous waste. Similarly, if a chemical's theoretical maximum concentration is one-tenth times its LC_{50}, then it could form an atmosphere one-tenth of its lethal atmosphere even with and would be considered a special under this classification proposal. The table below is used to classify the waste:

<u>Particulates:</u> Classification of a waste based on its particulate constituents would be based on the respirable fraction of the waste (PM_{10}, the fraction with a particle size less than 10 microns) times the sum of the ratios of each chemical's concentration (in mg/kg) in the respirable fraction divided by its inhalation LC_{50} (in mg/m3). This ratio accounts for the tendency of the chemical to be suspended in the air and for its acute toxicity by inhalation. DTSC proposes to classify a waste as hazardous if the sum of the concentrations of individual chemicals in the respirable fraction of the waste (in mg/kg) divided by their inhalation LC50s (in mg/m3) times the respirable fraction of the waste exceeds $2x10^6$, and to classify a waste as non-hazardous if the concentration of a chemical in the respirable fraction of the waste (in mg/kg) divided by its inhalation LC50 (in mg/m3) is less than 10^5. The rationale for these thresholds is as follows:

The concentration of a chemical in the waste multiplied by the particulate concentration in the air yields the airborne concentration of the chemical, assuming that the dust is suspended waste. Simplistically, if this airborne concentration exceeds the LC50 of the chemical, then the resulting concentration will be lethal. However not all airborne particles are respirable. If only a fraction of the waste is respirable, the expression must be corrected to account for the fraction of the waste that is respirable (F), and the concentration of the chemical (C) in the respirable fraction (PM10) of the waste must be used in place of the concentration in the waste as a whole. Thus, the expression for the lethal concentration becomes ($F * C * PM10 / LC50 > 1$). Because the endpoint is 50% lethality, a safety factor of ten is incorporated, making the expression F * C x

APPENDIX D: DTSC REPORT

PM10 / LC50 > 0.1. Finally, concentrations of the various toxic constituents in the waste must be added to determine the total toxic effect of the waste, i.e. $F * \Sigma(C \times PM10 / LC50 > 0.1)$. The two assumed airborne dust concentrations are based on the OSHA standard for respirable suspended particulates in the workplace (10^{-6} kg/m3) and the federal ambient air quality standard for PM10 (5×10^{-8} kg/m3). Thus, the exit threshold becomes $F * \Sigma(C \times 5 \times 10^{-8} / LC50 > 0.1)$, and the hazardous threshold becomes $F * \Sigma(C \times 10^{-6} / LC50 > 0.1)$. These expressions can be rearranged to give $F * \Sigma(C / LC50 > 2 \times 10^6)$, and $F * \Sigma(C \times / LC50 > 10^5)$, respectively

The following table is used to classify the waste:

Vapor Pressure/LC_{50} ratio sum	Classification	Concentration/LC_{50} sum
VP/LC < 0.1	Non-hazardous waste	C/LC_{50} < 10^5
0.1 < VP/LC < 1	Special Waste	na
VP/LC > 1	Hazardous Waste	C/LC_{50} > 10^5

Aquatic Toxicity

The current system classifies a waste as hazardous if its aquatic LC_{50} is less than 500 mg/l. DTSC proposes to classify a waste with an LC_{50} <500 mg/l as a special waste. A concentration of 500 mg/l would be equivalent to 7 tons in a two-acre lake five feet deep with complete mixing, which DTSC considers to be a reasonable worst-case release. As discused in apendix 2, above, a composite liner meeting RCRA Subtitle D specifications is assumed to reduce leakage from a landfill by 18-fold (tenth percentile estimate). Therefore, the waste could be 18 times as toxic (assuming that fish exposure is directly proportional to flow rate) without causing toxic effects on fish if it is placed in a subtitle D landfill. Therefore the proposed threshold for the fully hazardous tier is 500/18 = 30, i.e. a waste with an LC_{50} <30 mg/l would be classified as a hazardous waste